The Generativity Sciences

Robert R. Carkhuff

Published by: HRD Press, Inc.
                22 Amherst Road
                Amherst, MA 01002
                800-822-2801 (U.S. and Canada)
                413-253-3488
                413-253-3490 (fax)

ISBN 978-1-61014-381-3

Editorial services by Robert W. Carkhuff and Sally M. Farnham
Graphics and production by Jean S. Miller
Cover design by Eileen Klockars
Promotion by Swift Global Media

The Generativity Sciences

Contents

About the Author

Robert R. Carkhuff, Ph.D., is chairman, **The McLean Project.** One of the most-referenced scientists of the 20th century according to **The Institute for Scientific Research,** he is also recognized as **"The Father of Human Sciences"** due to his breakthroughs in our understanding of **Human Relating, Information Representing,** and **Human and Information Generativity.** Special recognition is due his paradigm-changing volumes on *The New Science of Possibilities* (2000) and its many applications (see References). Carkhuff's full body of work may be viewed on his websites:

www.carkhuff.com

www.mcleanproject.com

www.carkhuffgenerativitylibrary.com

Foreword*
"On These Sacred Grounds"

When Carkhuff met with the directors of his information technology (IT) and organization technology (OT) projects, he concluded: "We meet again on these sacred grounds."

What he was referring to was nature's sacred principles: nested multidimensionality, functional interdependency, generative processing."

Together, these principles constituted the core ingredients of Carkhuff's scientific mission, "The Nature of Nature":

- Interdependency of functions
- Multidimensionality of components
- Generativity of processes

On these "sacred grounds," Carkhuff believes, may be built an elevated civilization aligned with nature as a processing partner:

- peaceful with the interdependent generativity of cultural relating

- participative with the electronic generativity of enlightened governance

- prosperous with the entrepreneurial generativity of substance and wealth

Due to the "decomposition of all phenomena," Carkhuff believes "science is continuously and intentionally evolving." Carkhuff always concludes his "perspective pieces" with a transition. So here goes:

In transition, Carkhuff's work reflects the core processes of "Nature's Generativity":

"Carkhuff's contributions to Universal Processing, alone, qualify him for leadership among the greatest scientists of history. His 'nesting, encoding, and rotating of processing systems' are the core processes of 'Nature's Generativity.'

---

*Bernard G. Berenson, Ph.D. Keynote Address, *Nature's Generativity,* The American Noble Prize, January, 2011.

In creating The Human Sciences, Carkhuff's bodies of work belong in The Pantheon of Science along with the works of DaVinci, Newton, and Einstein."

Carkhuff's generativity, integrity, and productivity prove that we can indeed "become beams of light."

I still work indefatigably at science, but I have become an evil renegade who does not wish physics to be based on probabilities.

– Albert Einstein, 1948

I

Introduction and Overview

1 Science: The Voyage of Discovery

As applied social scientists, we may think of ourselves as intervening in our social environments—hopefully for the betterment of all of us.

As possibilities scientists, we may view ourselves as pioneers, building our cabins high up in the mountain ranges.

As generative scientists, we may position ourselves to view both the heavens above and the plains below, processing naturalistically and virtually to explore different objectives and test different routes to them. This is our "voyage of discovery."

The Phases of Discovery

We begin our voyage by ascending the mountains from the plains below. We take note of different phenomena along the way. We note that sometimes the phenomena appear almost simultaneously. That is to say, when one phenomenon appears, then another seems to occur. We see it in natural phenomena:

- Cumulus clouds bring rain.
- Rains bring plants.
- Plants become trees.

And so on…

We begin to build our images of our places within the phenomena. Most often, humankind has aspired to controlling the phenomena:

- Dams harness the water.
- Walls prevent flooding.
- Canals water the fields.

And so on…

We label scientists who think in terms of controlling nature probabilities scientists. They set up bell curves to assign all phenomena—natural and human—to their predicted places. They base their predictions on their descriptions of the phenomena. Thus, the functions of **Probabilities Sciences** and its conditioned populators may be viewed in Table 1.

Table 1. The Functions of Probabilities Sciences

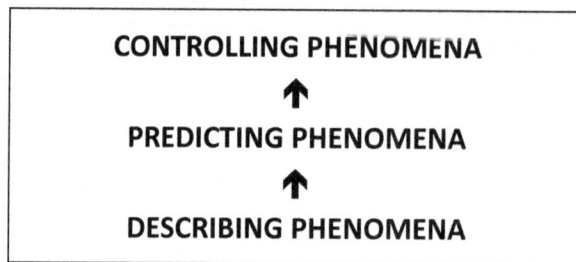

CONTROLLING PHENOMENA

↑

PREDICTING PHENOMENA

↑

DESCRIBING PHENOMENA

The problems of the **Probabilities Sciences** are with the waste of natural phenomena:

- The water is lost.
- The water is polluted.
- The lands are deserted.

And so on...

As it is with natural resources, so it is with social resources:

- People are lost.
- People are polluted.
- The environments are deserted.

And so on...

The Possibilities Sciences

There is another group of people who think in terms of aligning with all phenomena—natural and human. In effect, they treat others as they would be treated: relating to their experience and direction; empowering them to fulfill their potential; releasing or freeing them to define their own changeable destinies. Thus, the functions of **Possibilities Science** and its empowered participants may be viewed in Table 2.

Table 2.
The Functions of Possibilities Sciences

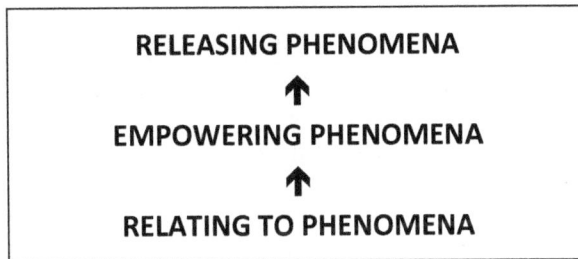

RELEASING PHENOMENA

↑

EMPOWERING PHENOMENA

↑

RELATING TO PHENOMENA

The benefits of the **Possibilities Sciences** are with the explosion of the potential of the natural phenomena:

- The water is naturally pure.
- The water is oxygenated to become pure.
- The water is life-giving in its purity.

And so on...

So it is with social resources:

- The people are naturally talented.
- The people's resources are actualized.
- The people's environments are optimized.

And so on...

The Generativity Sciences

There is still another group of people who generate "miracles of performance" in any area of natural or human endeavor. They are the people who live and process interdependently with all phenomena. They see no downside to the generativity of new and better ways of doing things. We label them "the generativity scientists." The sole focus—and "the soul focus" of everything they do—is illustrated in Table 3: **Freedom!**

Table 3. The Functions of Generativity Sciences

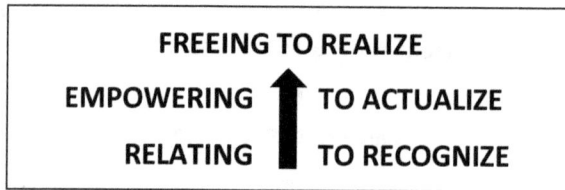

FREEING TO REALIZE		
EMPOWERING	↑	**TO ACTUALIZE**
RELATING		**TO RECOGNIZE**

- relating to recognize the potential of phenomenal resources
- empowering to actualize the potential of phenomenal resources
- freeing to realize the potential of phenomenal resources

These generators begin to build their cabins in the woods—their "chapels in the woods." They share their resources with the environment that surrounds them—the air, the water, the land. They position themselves to optimize their perspectives so that they can refine their objectives.

They give new definition to their objectives and build different pathways to them:

- to the rivers and waterways
- to the forests and hunting grounds
- to the towns that they establish with their models and their civilizations they build by their genesis

Gradually, they discover the best ways to successful human and natural achievements and the relations between them.

These processes are what generativity sciences are all about: discovering the best explanations of unknown phenomena so that we can discover the best ways of implementing shared needs:

- generating the operations to accomplish their purposes
- measuring the dimensions and effects of our purposes
- generating the total picture on maps that allow others to follow in our footsteps and then to draw them away

These are the ingredients of a relating and initiating science (see Table 4). Indeed, these are the ingredients of a growing and enduring civilization.

Table 4. The Processes of Generativity Sciences

- **EXPLAINING PHENOMENA**
- **DEFINING OPERATIONS**
- **MEASURING DIMENSIONS**
- **REPLICATING DISCOVERIES**

The scientific mission is to explain the mysteries of our universes—human and natural. In so doing, science may generate our human places in our natural worlds. At its peak, generativity sciences generate technologies that empower us to build a better world.

To do this, we must conquer the knowledge, skills, and attitudes of our times and then reject them to build a better place.

We must build our "chapels in the woods."

2 "Cubing"—The Vision of Information

One of the great lines from *Bambi* was when the deer's father said the following words: "Man is in the forest!"

When we, ourselves, initiate our "voyage of discovery," we might discover the products of earlier civilizations strewn among the relics of the forests:

- anything with a right angle like "squared"
- anything circular like wheels
- anything at all looking like a perfect replication of something else

And so on...

God did not make any phenomena perfectly equal to anything else. Not one cell! Not one atom! Not one subatomic particle!

Part of humankind's effort to assert its authority has been to exert its control. It is like replicating parts and their assemblage in a huge assembly-line and then insisting that this is the way God wants it!

Perhaps the most powerful images of human presence in the universes are the **"cubes"** that human brainpower generate. Cubing is humankind's way of representing the multidimensional nature of phenomena, and experience cubes are humankind's representation of the complexity of our universes—external and internal.

By definition, a cube is a solid bordered by six equal squares. The angle between any two adjacent faces is a right angle. Having defined it parametrically, we may refine it nonparametrically as a three-dimensional (3D) model of varying angles of interface. There is no requirement of solidity in generative processing; for the human's purposes, this 3D model may be related socially to other 3D models. In other words, the cubing of information serves humans by virtually representing their dimensions and their relationships. Cubing is humankind's viewing phenomena with the interior "third eye" just as we listen with our interior "third ear."

Cubing is the vision of information!

Information "Cubing"

The net of our processing of phenomenal experience may be viewed in our **Information "Cubing"** phenomena: phenomenal functions, components, and processes (see Figure 1). As may be noted:

- Phenomenal functions are driven by the discovery of new phenomena.
- Information components are driven by phenomenal representations.
- Human-information processes are driven by the phenomenal processing systems **(S–PP–R).**

We may summarize the result of **Information "Cubing"** operationally:

> **Phenomenal functions are achieved by information components empowered by human-information processing.**

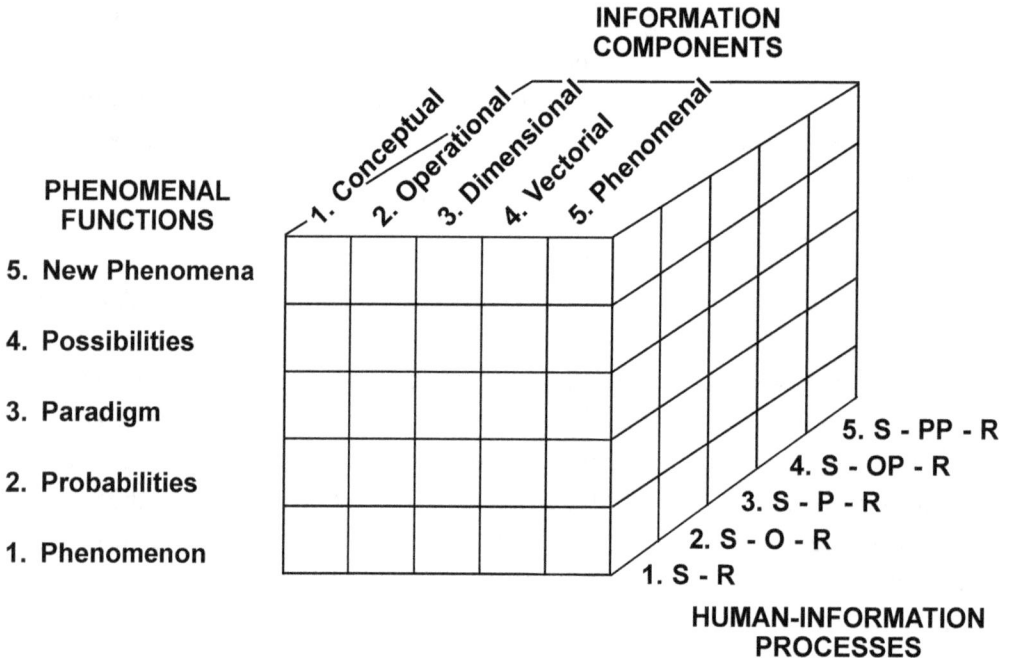

INFORMATION COMPONENTS

PHENOMENAL FUNCTIONS

1. Conceptual
2. Operational
3. Dimensional
4. Vectorial
5. Phenomenal

5. New Phenomena

4. Possibilities

3. Paradigm

2. Probabilities

1. Phenomenon

5. S - PP - R
4. S - OP - R
3. S - P - R
2. S - O - R
1. S - R

HUMAN-INFORMATION PROCESSES

Figure 1. "Cubing" Phenomena

We are now empowered to generate the phenomenal conditions within which the phenomenon is "nested" or "housed" (see Figure 2):

- **Generativity functions** are driven by freeing or releasing functions.
- **Phenomenal components** are driven by new phenomena.
- **Information processes** are driven by phenomenal images.

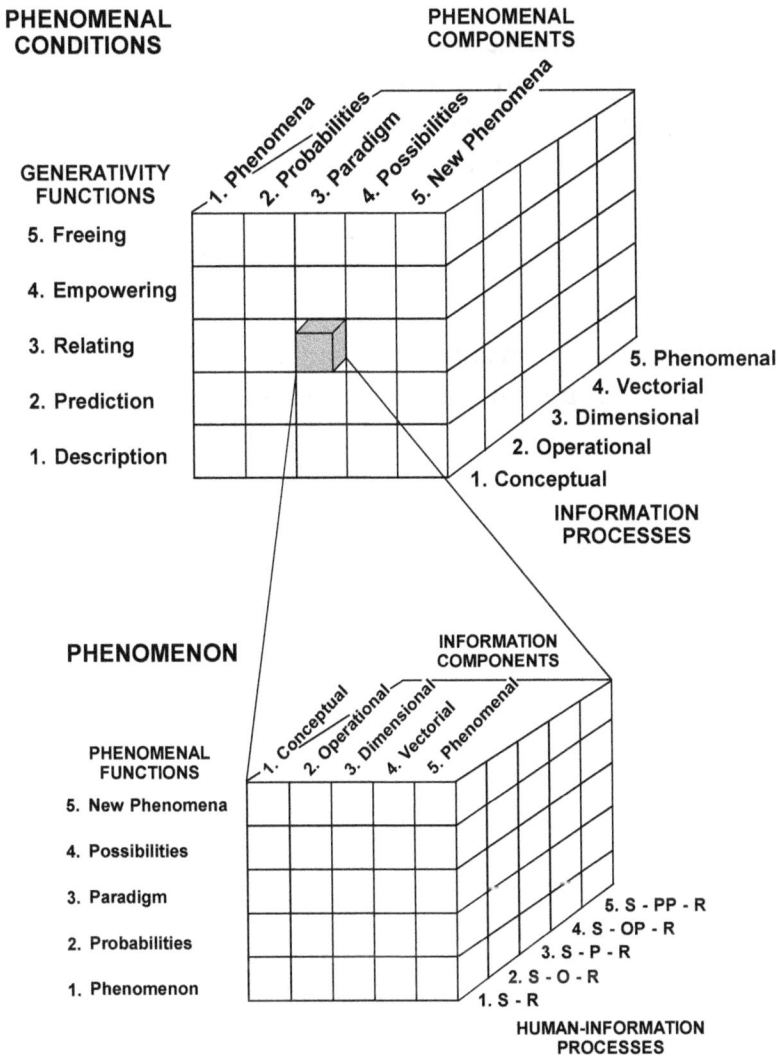

Figure 2. "Cubing" Phenomenal Conditions

Again, we may summarize the operational definition of conditions resulting from **Information "Cubing":**

> **Generativity functions are achieved by phenomenal components empowered by information processes.**

Finally, we can generate the phenomenal standards by which we measure our level of achievement (see Figure 3):

- **Information functions** are driven by phenomenal representations.
- **Human-information processing components** are driven by phenomenal processing systems **(S–PP–R).**
- **Mechanical processes** are driven by phenomenal performance standards.

Once again, we may summarize the operational definition of standards resulting from **"cubing":**

> **Information functions are achieved by human-information processing components empowered by mechanical processes.**

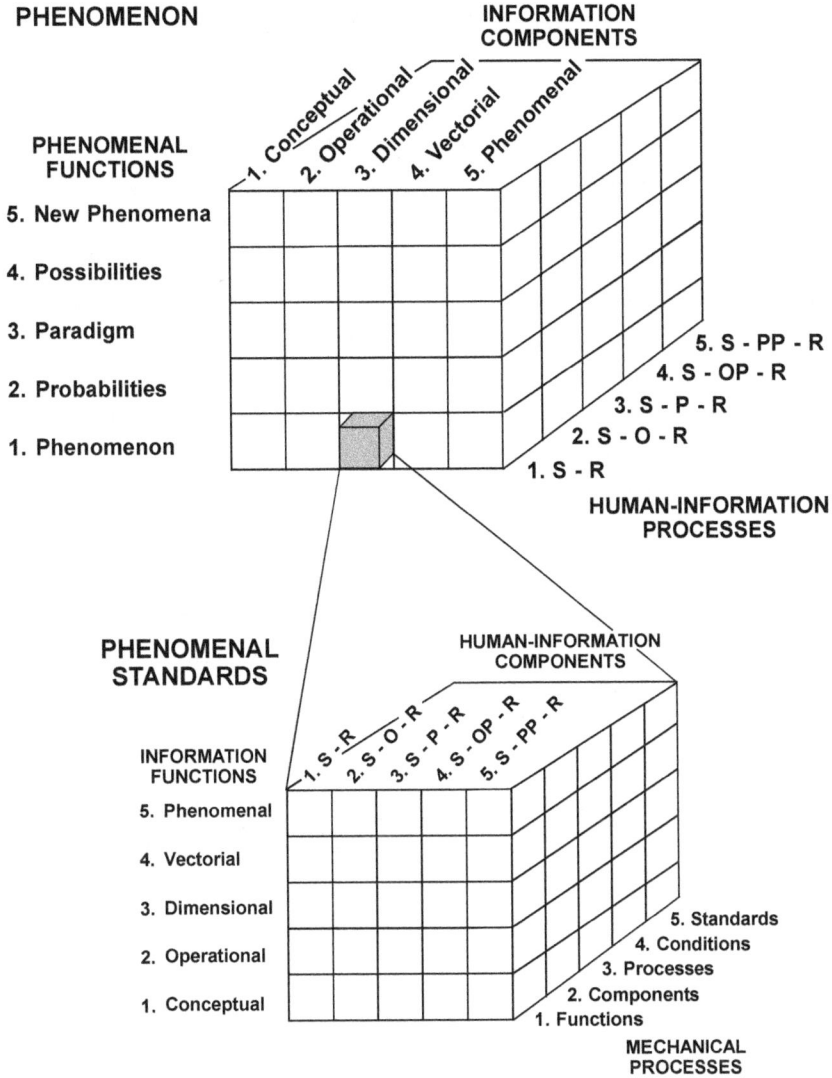

Figure 3. "Cubing" Phenomenal Standards

Information Is Living

The fundamental difference between living and nonliving things is information. Living things use information to birth, maintain, create, reproduce, and even terminate themselves. Nonliving things do not engage in any of these processes.

Information becomes potent when it is encoded in sequences of symbols such as alphabetical letters, musical notes, or some physical forms that are decoded by books and symphonies or some form of machinery. In short, information needs machinery that, in turn, needs information.

Here is how life works (see Figure 4). One chain of molecules is the DNA, which carries information. The other chain of amino acids is linked to proteins and carries on the business of life—birthing, growing, sustaining, reproducing. Notably DNA is not a "blueprint" but rather like a recipe with instructions to be followed.

In human cells, the DNA resides in 46 double-stranded chromosome chains. The chains are super-long and super-thin, reaching millions of miles, yet fitting into the nuclei of each of our trillions of cells.

"Cubing"—The Vision of Information

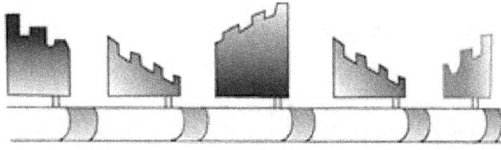

Information chains (DNA and RNA) made of four units (nucleotides)

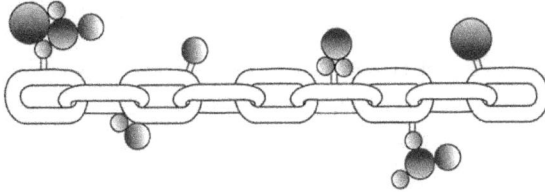

Working or structural chains (proteins) made of 20 units (amino acids)

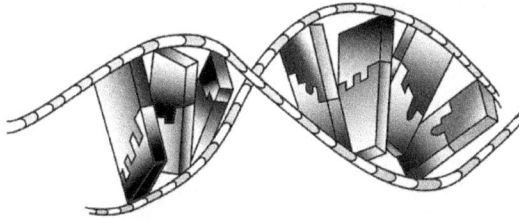

**For protection, easy access, and duplication,
DNA chains twist into a double helix.**

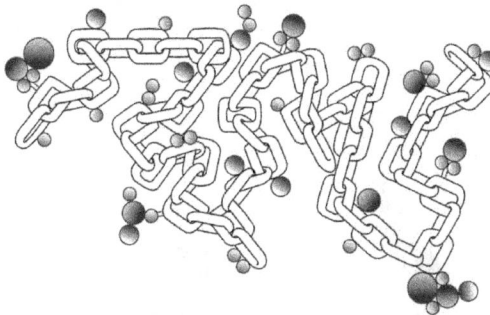

**For proper functioning, proteins fold into complicated shapes.
In this way, two-dimensional chains become three-dimensional machinery.**

Figure 4. How Life Works (Hoagland and Dodson, 1995)

The process by which the chains of molecules are empowered in a "nesting" of nucleotides (see Figure 5):

- **Nucleotides** are the smallest information units.
- **Genes** comprise nucleotides that specify proteins.
- **Chromosomes** are spooled strings of genes housed in a single unit.
- **Genomes** are collected in the nucleus of each of their cells.

The four nucleotides of **DNA** comprise the letters of its language of heredity:

- **A** – Adenylic acid
- **T** – Thymidylic acid
- **C** – Cytidylic acid
- **G** – Gaunylic acid

Each is a unique arrangement of carbon, nitrogen, oxygen, and hydrogenations called a base.

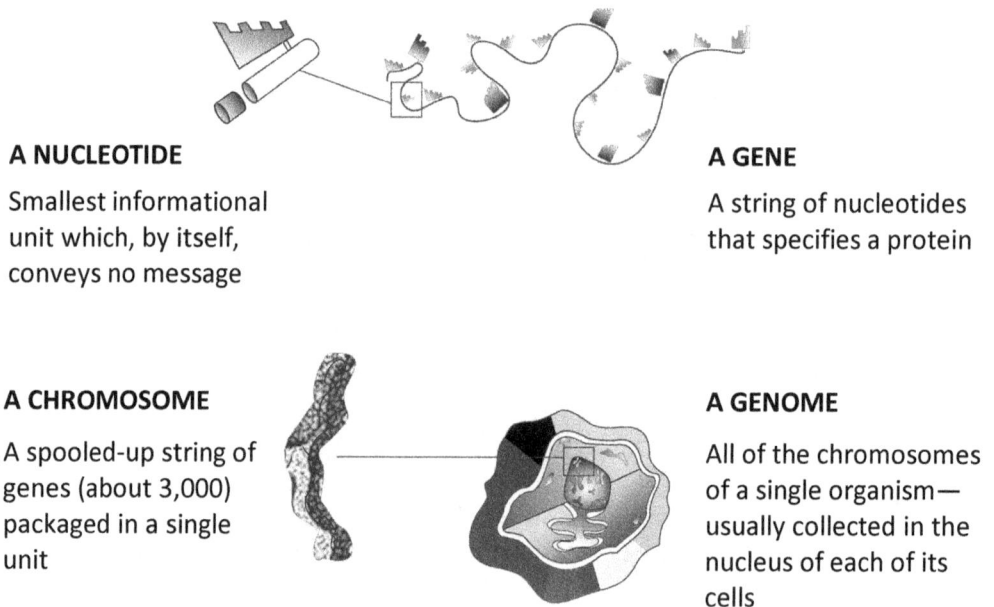

A NUCLEOTIDE

Smallest informational unit which, by itself, conveys no message

A GENE

A string of nucleotides that specifies a protein

A CHROMOSOME

A spooled-up string of genes (about 3,000) packaged in a single unit

A GENOME

All of the chromosomes of a single organism—usually collected in the nucleus of each of its cells

Figure 5. The "Nesting" of Nucleotides

"Cubing" Life

Just as we "cube" all other phenomena in human experience and human endeavors, we may "cube" DNA information (see Figure 6). As may be viewed, the following phenomena are represented:

- **Life functions**
- **DNA components**
- **Nucleotidal processes**

We may define its operational definition as follows:

> **Life functions are achieved by DNA components empowered by Nucleotidal processes.**

Cubing these life operations empowers human generativity. The human is not simply the dependent recipient; the human becomes a collaborative partner with the very life experiences that are impacting him or her. The human becomes the interdependent and synergistic generator with life itself: each grows as the other grows.

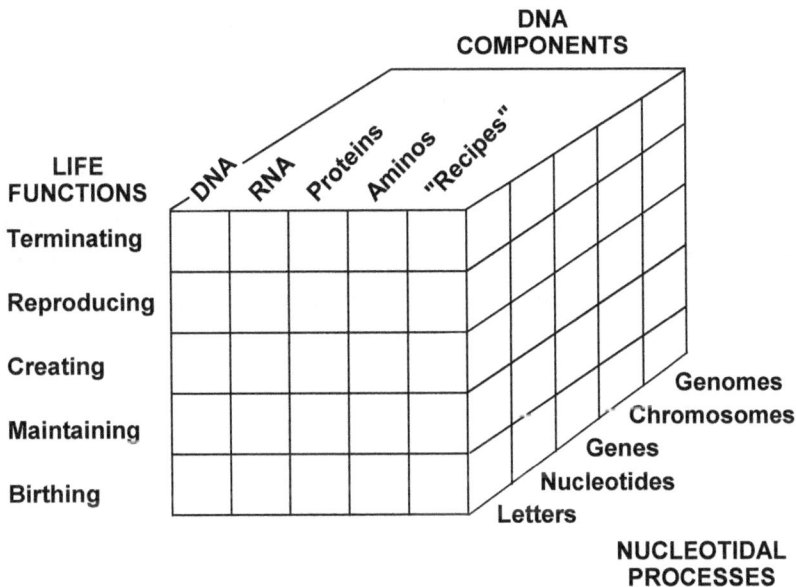

Figure 6. "Cubing" DNA Phenomenal Information

We can generate the very source of DNA by rotating the DNA phenomenon inductively or clockwise (see Figure 7). As may be noted, the life components are now dedicated to the generativity functions and empowered by the DNA processes.

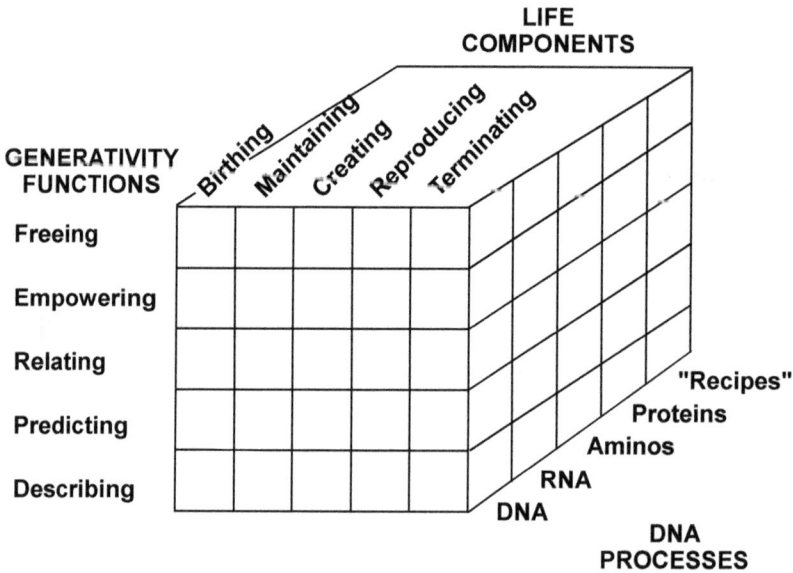

Figure 7. "Cubing" DNA Phenomenal Conditions

Similarly, we can generate the performance standards of DNA by rotating DNA phenomena deductively or counterclockwise (see Figure 8). As may be noted, the Nucleotide components are now dedicated to DNA functions and empowered by mechanical processes: functions, components, processes, conditions, standards.

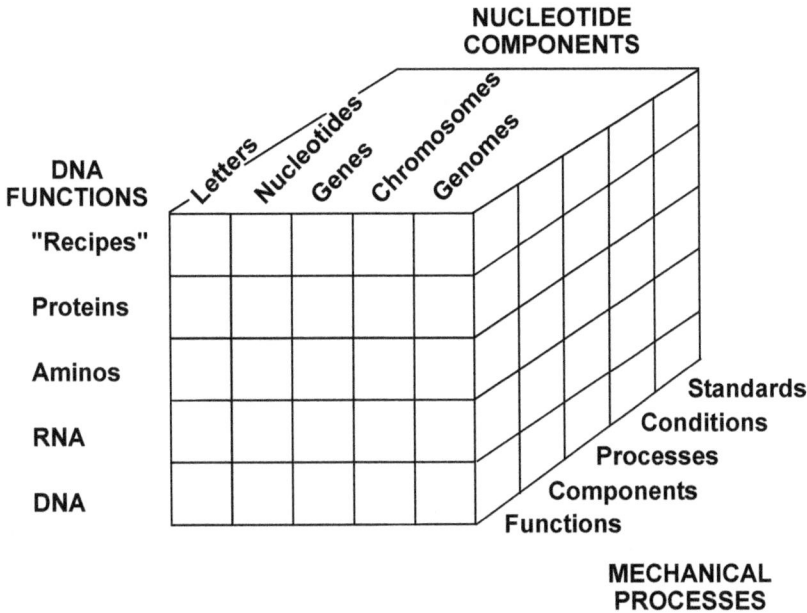

Figure 8. "Cubing" Performance Standards

As may be noted, the human processors may induce the very generativity that will guide them through life. By inductively rotating (clockwise) to the generativity functions to which the life components were dedicated, the human processors are empowered to generate their life's missions:

- to relate interdependently
- to empower intentionally
- to free intellectually

Now we can ask the critical questions of life:

- How can we become synergistic with life's experiences?
- What are the resource and skill level requirements?
- What are the degrees of freedom and enterprise required?

Modeled on DNA, Information "Cubing" empowers us to model all of life's operations, measure our performance, and project our future opportunities.

Information "Cubing" is the source of life itself!

II

The Methods of Sciences

3 Probabilities—The Probabilities Sciences

As we initiated our "voyage of discovery," we may generate visions of the environment in the mountains, rivers, and plains. Generally, we draw these in our mind's eye as undulating waves and static deserts.

We also draw virtual representations of the cabin in which we would live. Generally, these images are cubist in nature, with right angles everywhere. We place these images in the middle of our virtual environment. That way, we are positioned to view all natural phenomena. Moreover, we are positioned to note the changes in the phenomena.

What we cannot do well was represent the people in our environment. Indeed, if there were people, what would we represent? After long thought, we may conclude that what would be capital or most important about these people would be their ways of thinking and the images or products they produce:

- Could they represent the mountains, plains, and rivers with differentiated but undulating waves?

- Could they represent their homes or workplaces with **3D "cubing"**?

- Could they represent strangers such as us with products of our processing? Or would we seem like alien creatures from another planet? It makes all the difference in how we relate together or deteriorate apart!

Fast forward 60 years to what I am today: I found the Probabilities Sciences with which I was indoctrinated to be sufficient for all definable human circumstances. Indeed, the Probabilities Sciences dominated human existence and defined the conditions of our scientific experiments. To be sure, for 14 million years, since the first signs of hominid life, humankind and nature have related to each other through one processing system: **S–R Conditioning** or reflex responding where there is no intervening processing between the presentation of the **stimuli (S)** and the emittance of a **response (R)**.

Initiating Probabilistically

In this context, Parametric Science evolved in the 20th century. It defined the parametrics of phenomena and projected probabilities for their occurrence. For example, the entire area of parametric statistics evolved from agrarian applications: rows and columns of agriculture were treated differentially. The deviations or variability of phenomenal distributions around some estimate of central tendency were selected to meet environmental or marketplace requirements. Thus, the controlling function of science was enabled by the describing and predicting functions. The Probabilities Science of today was defined by the Parametric Models of Measurement.

	FUNCTIONS	PROCESSING	MEASUREMENT
PROBABILITIES SCIENCE	• Control • Predict • Describe	S–R Conditioned Responding Systems	Parametric Measurement

The contributions of Gauss, Pascal, and others to Probabilities Science have enabled all processors to construct a world of Probabilities Mathematics and Normal Curves. They have empowered us to start our "voyages of discovery." The problem is this: They are inadequate in the manner that they pursue the mission of science, "The Explication of the Unknown."

In the simplest terms: God has never made any species of life equal in their operations, let alone symmetrical distributions. Only with man-made things do humans seek to impose statistical process controls with equal-interval measurements in order to narrow the tolerances of his tools and products!

To be sure, one of the "miracles" of Probabilities Science was "The Normal Curve" (see Figure 9). The Normal Curve is a unimodal frequency distribution curve. Its scores are plotted on the x-axis and their frequency of occurrence on the y-axis:

"It is the lifeblood of descriptive statistics" for any characteristics of any phenomena, human or otherwise. (Sprinthall 2003)

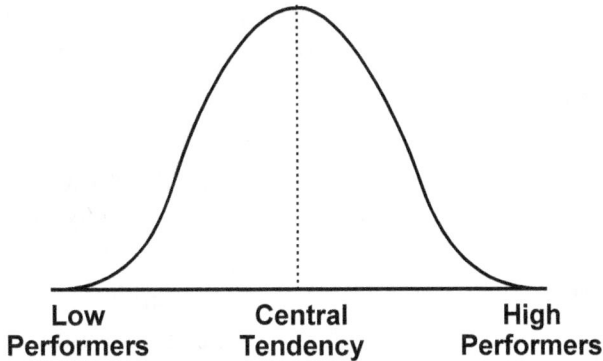

**Low
Performers**   **Central
Tendency**   **High
Performers**

Figure 9. The Normal Distribution of Parametric Measurement

The Normal Curve serves many functions (see Figure 10). Among them are the following:

1. The central tendencies converge to reduce the variability beyond these averages.

2. Because of the symmetry of the curve, precisely 34.13% of the scores fall within one standard deviation of the mean.

3. As we move farther from the mean, a declining number of scores fall within two (13.59%) and three (2.15%) units of standard deviation for a grand total of 99.74% of the scores.

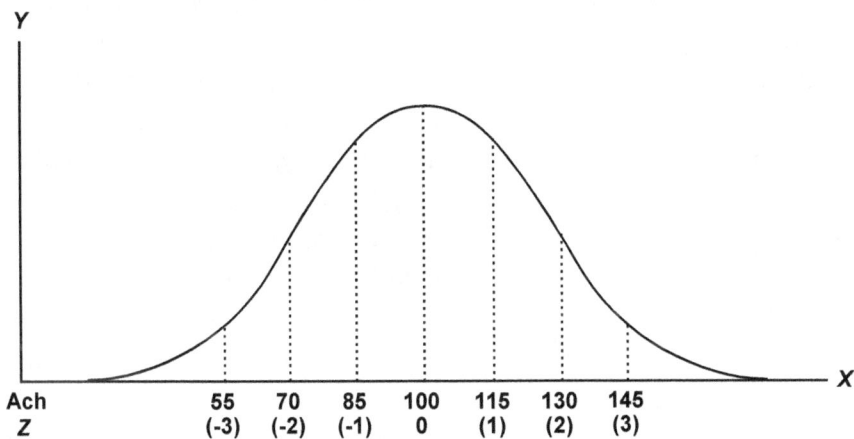

Figure 10. The Normal Distribution of Measurements

It is this precision in description that makes the Normal Curve powerful in the interpretation of raw score performance. Succinctly, it takes into account the central tendencies, the variability, and the standard deviation. This empowers us to gain an understanding of individual phenomenal performance relative to the phenomenal performance of others in the reference group. Moreover, we can compare individual performance on two or more normally distributed scores.

As useful as the normal distribution is for comparing performance, it is also a powerful source of controlling performance. Among other controls are the following (see Table 5):

Table 5. Rules for "Controlling" Probabilities

Levels	Operations
5	Controlling Development of New Curves
4	Reinforcing Movement Outside of Curves
3	Reinforcing Movement within Curve
2	Predicting Placement in Curve
1	Defining Normal Curve

The Normal Curve, while defining the parameters of characteristics such as intelligence or performance, also limits the reference group's standards for achievement. Furthermore, it is used to reinforce individual movement both within and outside the curve of achievement, thus limiting aspirations or intentions for achievement. Finally, it directly controls the development of new curves, thus limiting the progress to achievement in new areas of performance, such as new areas of intelligence.

In summary, while the Normal Curve fulfills descriptive and predictive requirements, it also "controls" the standards of achievement for the people or phenomena being addressed. For example, it tends to limit our thinking to best practices in all areas of human endeavor. Indeed, often it is not until people are forced by the exemplary performance of others that they create new visions and missions and thus new curves for performance.

Probabilities Science gives us an image of the probable. In our work, we label this **"The Current Operating Paradigm"** or **"COP."** Because we seek **"The Productive Operating Paradigm"** or **"POP,"** we label the first a **"COP-OUT."** It may be the best practice, but it is never the best idea!

The "net" of Probabilities Science and Parametric Measurement is that it is a "heuristic" place to start processing.[*] It provides a structure that constitutes the processing threshold of all sciences and all processing. It is itself a hypothetical construct of reality calculated to initiate processing.

---

[*] For students of the problems of unequal variance that follow, there are solutions. In order to statistically verify the assumption of equal variability, the test most commonly used is Levene (1960). It can be used for both *t* tests and *F* ratios. It is basically a variation on ANOVA where the scores are presented as deviation scores rather than scores in their original form. Students may search our Levene's "Robust tests for equality of variances" in R. C. Sprinthall, *Basic Statistical* Analysis (Boston, MA: Allyn and Bacon, 2012, p. 593 ff.).

4 Possibilities—The Possibilities Sciences

As we continue on our "journey of discovery," we may learn that much of the world is not what we have been told. The operations are different; the measurements are complex; the maps cannot be followed.

- One of the mountains dwarfs the other mountains: all of the other mountains exist in its shadow.

- The rivers explode savagely, free of the canals that drain the power of rivers in civilized territories.

- The humans we meet are free of our civilized behaviors but eager to know our intentions.

And so on...

In short, everything seems outside of the limiting parameters that have been drawn for us. We need another way of representing them accurately and truthfully.

Fast forward to our time: as useful as the parametric distributions were for comparisons of performance, these are also a powerful source of controlling performance. As we encountered human experiences and endeavors, we found the Probabilities Sciences limited in their abilities to describe and predict human performance, let alone to control it. Indeed, it was precisely the introduction of the Data Age that led to the Possibilities Science. All of the branching, parallel, and interactive processing systems of **Information Technology** or **IT** are based upon **S–O–R Discriminative Learning Systems.** They require extraordinary repertoires of conditioned responses with which the human **organism (O)** discriminates the **stimuli (S)** and emits the appropriate **responses (R)**.

Possibilities Science is the source of changeability: process-centricity, the continuous processing of all phenomenal dimensions. As such, it expands 360° of global possibilities of continuously changing, interdependent, and asymmetrically curvilinear multidimensional phenomenal vectors. These phenomenal possibilities are due to the processing ability to align with the phenomena in their naturalistic form. In this context, probabilities phenomena occupy a small window of opportunity in space and time of the "possibilities universes" as "probabilities space-times."

Possibilities Science addresses the limitations of the Probabilities Model. The critical difference between possibilities and probabilities is found in the quality of the dimensions and their interdependent relationships in continuously evolving processing systems. The operative words are **"processing systems":** all dimensions are defined as processing systems. Specifically, the culminating function is to **release or free** the changeability in phenomena. This requires the highest level of phenomenal information: levels that enable us to **relate** and **empower** phenomena in order to **release** them.

SCIENCES	FUNCTIONS	PROCESSING	MEASUREMENT
POSSIBILITIES	• Release • Empower • Relate	S–O–R Discriminative Processing Systems	Non-Parametric Measurement

Ironically, it is with the introduction of Possibilities Science that the Normal Curve and Parametric Measurement come to life. Indeed, the Possibilities Scientist "anthropomorphizes" the phenomena being studied.

Possibilities Scientists bring the phenomena to life by aligning accurately with their external operations and then merging empathically with their internal operations. It is as if the scientist becomes one with the phenomenon.

Under the principles of Possibilities Science, hypothetical–deductive model-building is the source of this science of change: probabilistic models are built inductively and possibilistic hypotheses are derived deductively (see Figure 11):

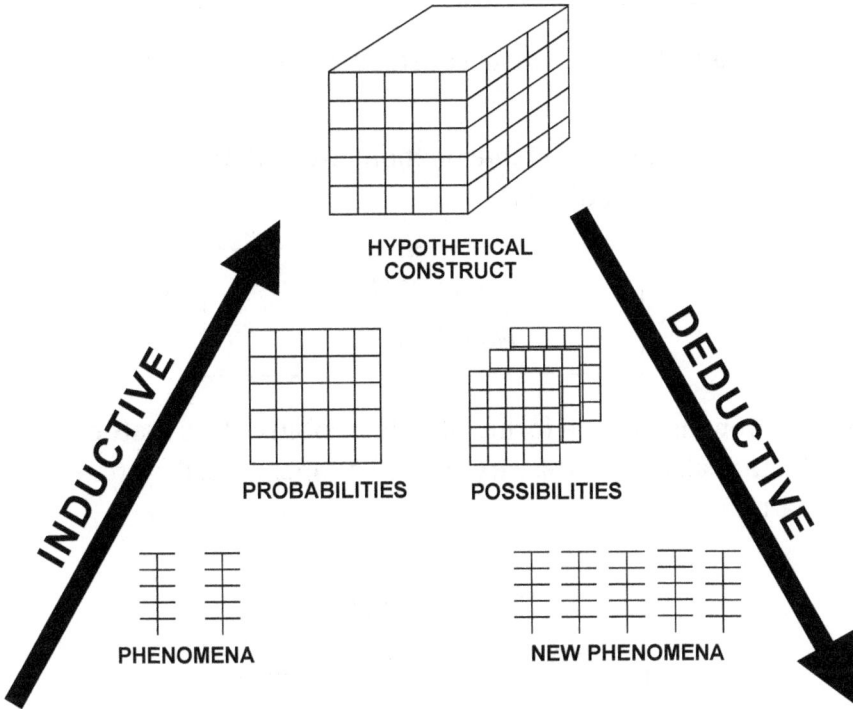

Figure 11. The Building and Testing of Hypothetical Constructs

As may be viewed, hypothetical constructs are built inductively as follows:

- The observations of stable phenomena that are factored and scaled in related dimensions:

> | |
> |---|
> | *Components* |
> | *Functions* |

- The dimensions are formulated in **probabilities statements** (historically laws) that are related in operational systems and matrices:

> | | |
> |---|---|
> | **If** | *Components* |
> | **Then** | *Functions* |

- The probabilities dimensions are then related in hypothetical constructs (historically theories) that are related multidimensionally in schematics and models:

> | | |
> |---|---|
> | **If** | *Components* |
> | **By** | *Processes* |
> | **Then** | *Functions* |

In turn, the hypothetical constructs are the source of systematic deductive hypothesis testing:

- The **possibilities statements** are deduced from the hypothetical constructs occurring under differing phenomenal conditions (historically theorems) that are formulated as vectorial relationships:

Under	*Conditions*
If	*Components*
By	*Processes*
Then	*Functions*

- The hypotheses are deduced from the possibilities statements and tested empirically for achievement of standards under differing phenomenal conditions resulting in new phenomena:

Under	*Conditions*
If	*Components*
By	*Processes*
Then	*Functions*
So that	*Standards*

This Hypothetical–Deductive Model-Building and Testing continues with the generation of new phenomena in replacement of the old phenomena: observations, probabilities, constructs, possibilities, hypotheses. Together, they culminate in the generation of New Hypothetical–Deductive Model-Building and Testing.

We may summarize our model building with a **Verbal Articulation of Phenomena** (see Figure 12).

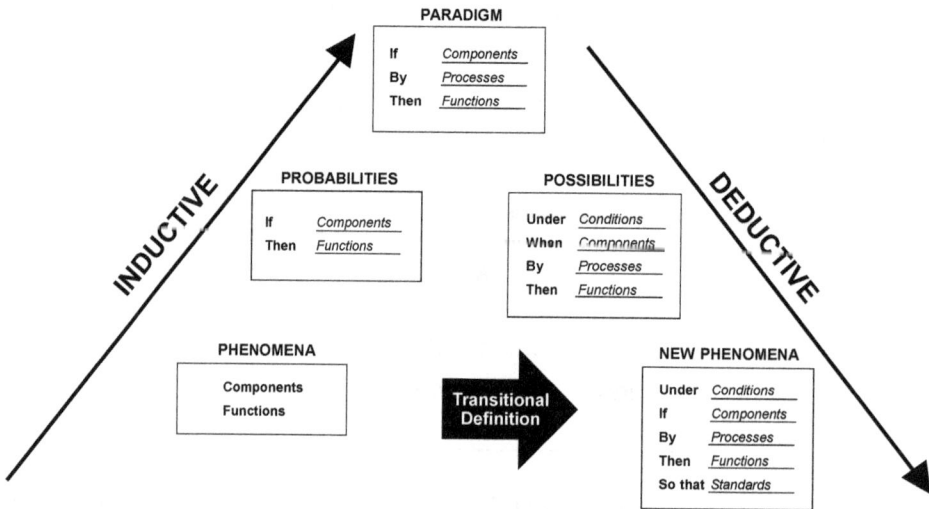

Figure 12. Verbal Articulation of Phenomena

As may be noted, the scientific models are built inductively and the technological hypotheses are derived deductively. This means that the models or paradigms are based upon inductive observations as **probabilities statements.** Concurrently, it means that the technologies are based upon deductive derivations as **possibilities statements.** As we will soon see, every application or transfer of a technology is a testable hypothesis.

The Best Idea or exemplary performance is introduced by focusing upon its placement beyond six sigmas of the Probabilities Curve. It moves progressively through the operations of changeability that define the New Possibilities Curve (see Table 6):

Table 6. The Progressive Movement
Toward Changeability

Levels	Operations
5	Enhancing
4	Embodying
3	Incorporating
2	Approaching
1	Focusing

We may view the lifeful movement of the curve in Figure 13 on the following page. As the high performer or the best idea attracts attention, the curve begins to move like a humpback whale toward higher performance.

The takeaway from Possibilities Science and Non-Parametric Measurement is the movement from the Controlling Function of Probabilities to the Releasing Function of Possibilities. It is the beginning of a joyous "voyage of discovery" toward generativity that defines the "best processes for generating the best ideas."

The Generativity Sciences

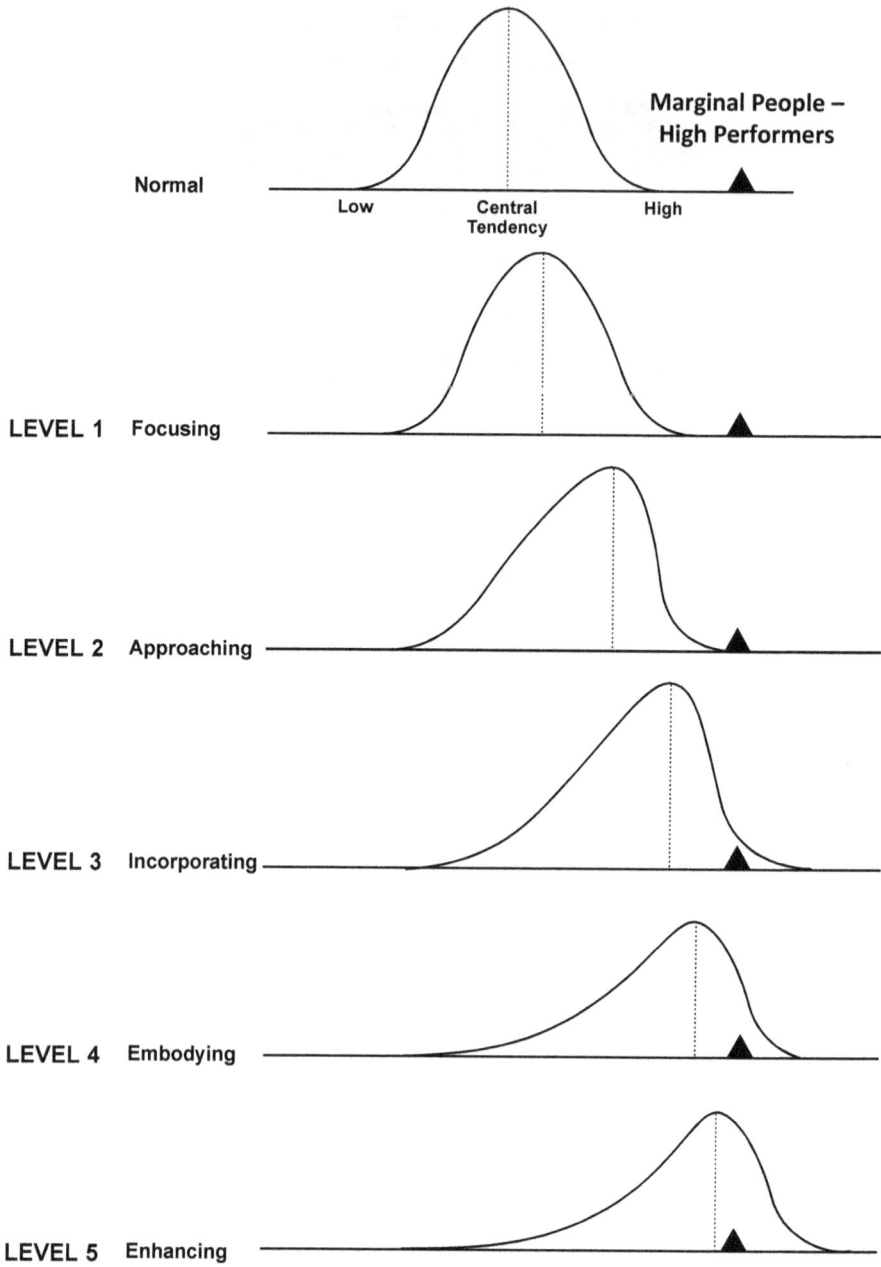

Marginal People –
High Performers

Normal
Low   Central   High
Tendency

LEVEL 1   Focusing

LEVEL 2   **Approaching**

LEVEL 3   Incorporating

LEVEL 4   Embodying

LEVEL 5   Enhancing

**Figure 13. The Progressive Movement of Transforming
Possibilistically**

5 Generativity—The Generativity Sciences

Finally, we may find ourselves in our living place and working space. Our experiences with the possibilities of life impacted us to consider a life of change rather than a life of stasis. Change could be generated, but it was outside the parameters set for us. It carried with it great risks, but also potentially great rewards:

- We could "position" ourselves at the top of the highest mountains and embrace its view of nearly everything around us.

- Or we could seek the safety of living in the shadows of this mountain, protected from the fury of violent changes in the weather and the geophysics of the environment.

- We could prepare ourselves for either place by changing our very nature and behavior.

And so on...

Fast forward to the perspective I had now expanded. As the universes of possibilities expand, they generate "possibilities spaces" or "phenomenal spaces," opportunities in which Generativity Science may work its magic. Generativity Science is an outgrowth of Possibilities Science. Dedicated exclusively to the Freedom Function, Generativity Science emphasizes generative human processing of information-modeling components: conceptual, operational, dimensional, vectorial, phenomenal. In other words, Generativity Science emphasizes expanding the levels of information-modeling before narrowing the objectives by scaling the values and requirements generated by the decision-makers, human or phenomenal.

It is with the introduction of Generativity Science that the Possibilities Science is culminated. To be sure, the Possibilities Scientist, himself released by Generativity, seeks to actualize the freedom function for all phenomena. Generativity empowers both scientist and phenomena to generate life on their own terms. Generativity is not simply "thinking outside of the box." Generativity is creating the "cubes" and "spheres" and "social schematics" to fill the possibilities spaces. Generativity is creating our own "universes—

internal and external"—in which we and the phenomena we are addressing live.

SCIENCES	FUNCTION	PROCESSING	MEASUREMENT
GENERATIVITY	Free	S–P–R Generative Processing Systems	Paradigmetric

We may illustrate by operationalizing just *one* of these processing systems with all of the information functions (see Figure 14).

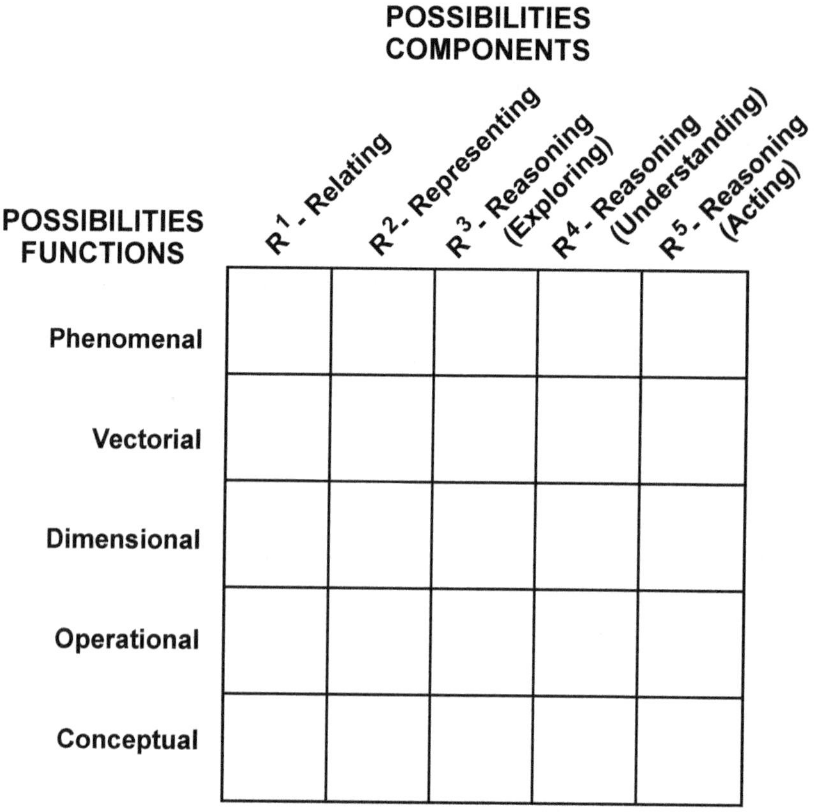

POSSIBILITIES COMPONENTS

Figure 14. S–P–R Generating with Phenomenal Information

As may be noted, **S–P–R** generativity processes possibilities information functions as follows:

- R^1 – **Relating** to data
- R^2 – **Representing** information
- R^3 – **Reasoning** by exploring information
- R^4 – **Reasoning** by understanding information
- R^5 – **Reasoning** by acting upon information

The levels of elevated processing generate elevated levels of information.

One generativity system, "The Human Generative Processing System," is illustrated on the following page (see Figure 15). It illustrates "thinking beyond the high-beams." It is as if we were driving along as we use our brainpower in most of our lives—using our low-beams. Once in a while at night, we switch to high-beams and see things that we would never have otherwise seen. This enables us not only to avoid accidents with the proba-bilistic obstacles lying ahead of us, but also to get better perspectives on our possibilistic targets or goals.

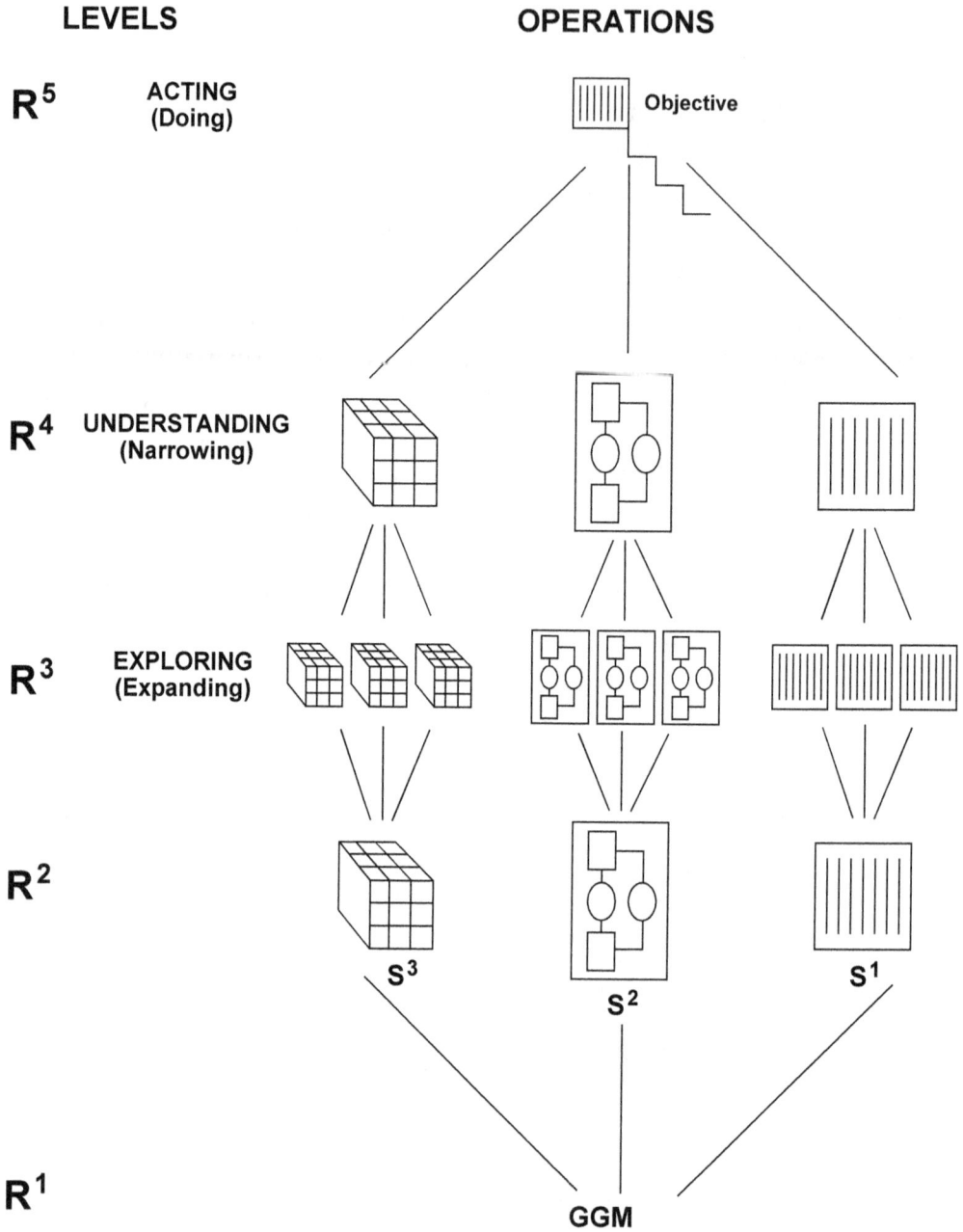

Figure 15. Generative Processing Systems

We may take a simple overview of the structure and functions of generativity operations. As may be seen, the phenomena is a simple one. It begins with a simple image of information input or knowledge or data:

It ends with a concrete image of information output, or an operational action plan:

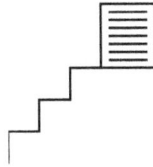

Generativity goes through five simple processing operations to transform the information input into action output:

- **R^1 – Relating** to images of information
- **R^2 – Representing** operational images
- **R^3 – Reasoning** by expanding images
- **R^4 – Reasoning** by narrowing images
- **R^5 – Reasoning** by acting upon images

That is it! Five simple steps. They meet the criteria of science. They are operational and therefore doable. They are replicable and therefore repeatable. They are measurable and therefore add to our storehouse of scientific knowledge. Moreover, they are elegant and therefore highly leveraged: they require a minimum amount of explanation to account for a maximum amount of effect.

When we flick on our "thinking beams," we see beyond the high-beams. This empowers us to not only see things that lie before us—probabilities. It also empowers us to see things that might be—in other words, possibilities. This empowers us to view the possibilities in our lives. Even in the face of failure, there are 360° global degrees of success experiences "looming" ahead.

We may view the generative movement of the curve in Figure 16. Each curve eclipses the performance of the previous curve as low performers are transformed into high performers by **relating, representing,** and **reasoning** (see Table 7).

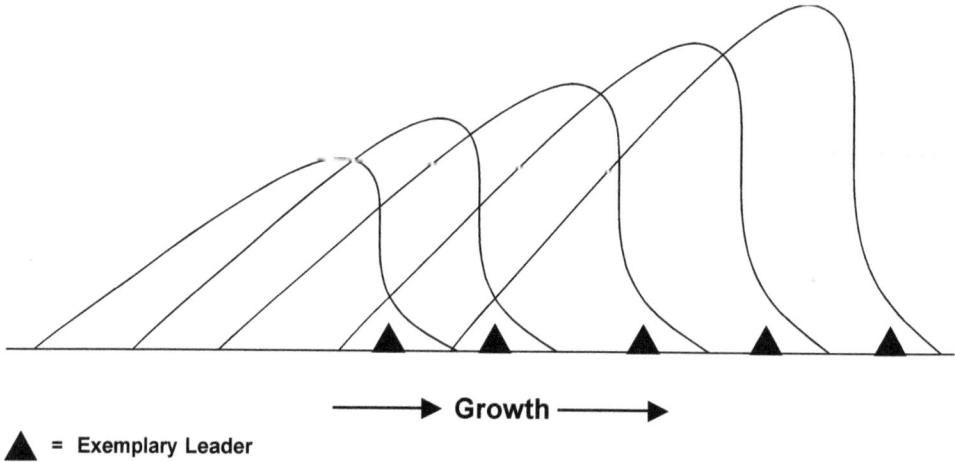

➤ Growth ➤

▲ = Exemplary Leader

Figure 16. Exemplary Leadership and Continuous Growth

Table 7. The Culminating Movement of Changeability

Levels	Operations
5	Reasoning by Acting
4	Reasoning by Understanding
3	Reasoning by Exploring
2	Representing
1	Relating

As was noted earlier in Generativity Science, the continuously elevating curves "follow the leaders" whether internal or external. For example, Figure 17 illustrates the real-world movement of economic growth as a function of American Free Enterprise Leadership over the past five decades.

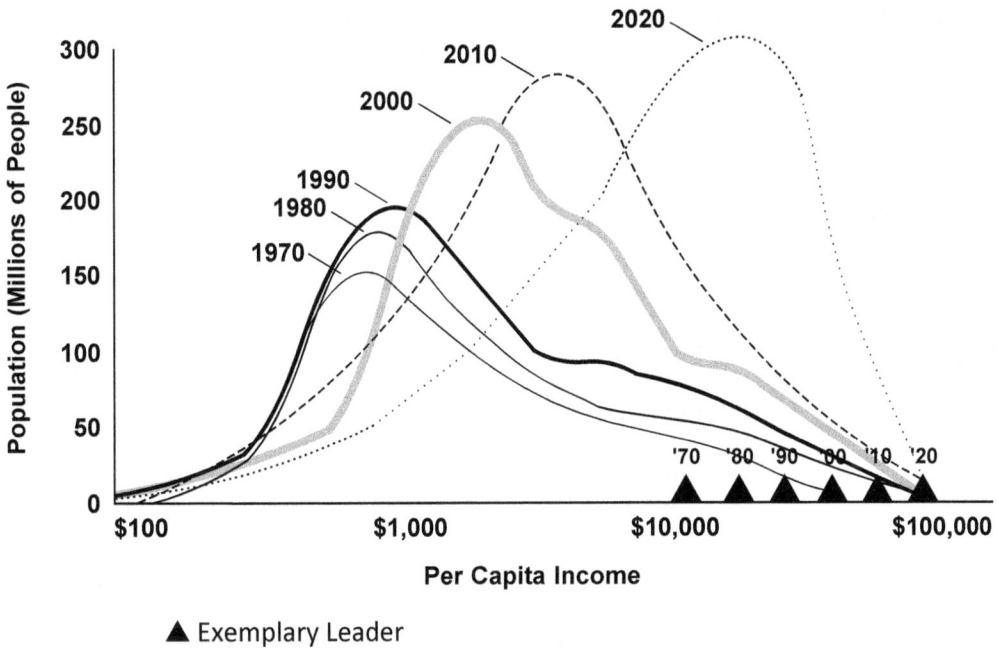

▲ Exemplary Leader

Figure 17. Exemplary U.S. Leadership and World Per Capita Growth (by Decades)

In conclusion, the culminating benefit of Generativity Science is that individuals, organizations, communities, cultures, and nations may continuously define and redefine their own destinies. Built upon the Theory of Relativity, the Science of Generativity culminates the Science of Change.

Together, the sciences comprise the Science of Change. They constitute all of the sciences that we can draw upon to accomplish any human missions that we seek:

- **The Probabilities Science** provides the distribution of predictable responses with projected costs and benefits.

- **The Possibilities Science** expands the alternative responses with leveraged costs and elevated benefits.

- **The Generativity Science** generates the new responses with potentially diminishing costs and accelerating benefits.

THE SCIENCE OF CHANGE

SCIENCES	FUNCTIONS	PROCESSING	MEASUREMENT
GENERATIVITY	• Free	S–P–R	Paradigmetric
POSSIBILITIES	• Release • Empower • Relate	S–O–R	Non-Parametric
PROBABILITIES	• Control • Predict • Describe	S–R	Parametric

The Science of Change empowers us to generate solutions to any problems in any human endeavor. Together, they have provided us with a roadmap that is developmental and cumulative:

- We initiate probabilistically.
- We transform possibilistically.
- We culminate generatively.

The Science of Change has evolved intentionally in the 21st century. It empowers us with the capacity to generate our own solutions to our own human problems.

6 Intentionality—The Science of Change

Now we are matured—psychologically and intellectually:

- with the support of Probabilities Sciences and parametric measurement
- with the perspective of Possibilities Sciences and non-parametric measurement
- with the initiative of Generativity Sciences and paradigmetric measurement

We have tolerance for the parametrics of **probabilities traditions.** There was once a reason for everything they do.

We embrace the non-parametrics of **possibilities experiences.** We must conquer them before we judge them.

We have a love for the paradigmetrics of **existential generativity.** We are released to go anywhere and become anyone. We are free at last!

Free at last!

Until science began to generate its own history, the maturity of science shadowed the history of civilization. Beginning with its own philosophical base, science reflected the sensory experience of the philosopher–scientists and presented anecdotal evidence as its facsimile of descriptive measurement. Continuing its evolution by studying the phenomena, themselves, the phenomenal–scientists presented empirical evidence in support of their testable hypotheses and predictive measurement.

The processing threshold of science was crossed with entry into Probabilities Science and the discovery of parametric measurement. Parametrics yielded systematic ways of describing and predicting all phenomena in terms of the distribution of their measurements. Unfortunately, these measurements also controlled our evolving visions of the phenomena studied.

Rather than controlling phenomena, Possibilities Science aligned with the natural response dispositions of the phenomena. In so doing, possibilities sought non-parametric measurement for its supporting bodies of evidence. In replacement of control, possibilities sought to *relate* to the potential operations of the phenomena, **initiate** to *empower* that potential, and *release* or *free* the phenomena with matured potential to define their own destinies in the marketplaces of the world.

Finally, the Generativity Science sought to actualize the freedom of the phenomena by intervening to empower their processing potential to higher and higher levels. In so doing, it used for measurement the very same paradigmetric measurements that it had employed to generate the original paradigms. In replacement of releasing empowered phenomena, generativity sought to *relate* the brainpower to their information worlds, *represent* the information paradigmetrically, and *reason* to explore, understand, and act upon the now-productive information.

Everything we do in life involves scaling. At the extreme, it may be scaling ourselves from "where we are" to "where we want or need to be." In sophisticated terms, it may be developing a matrix of two scales: "What are my requirements?" and "What are my capacities to meet these requirements?" This may lead to super-sophisticated terms of measuring our skills, knowledge, and attitudes in relation to the requirements of an organization to which we are applying for schooling or working. We could go on. Everything is scaling! Everything!

Even science reduces to scales. Historically, there are many different kinds of scales. Dominant among traditional scales are the following: nominal or naming; ordinal or numbering; and interval or measuring by equal intervals. None of these suffice for the breakthrough architectures that we require to generate "new paradigms for change."

Now there are paradigmetric scales. Paradigmetric scales are drawn from the paradigms or models we generate. We may first develop the scales inductively, and put them together in various combinations. Or, conversely, we may generate conceptual models and then deductively scale the dimensions of these models. In either event, we require paradigmetric scales to measure our effectiveness in both building and applying our paradigms (see Table 8).

Table 8. Rules for "Scaling" Paradigmetrically

Levels	Operations
5	Culminating Mission (Destination)
4	Cumulative Processing (Collaboration)
3	Processing Threshold (Generativity)
2	Developmental Steps (Discrimination)
1	Baseline (Origin)

The rules for paradigmetric scaling are the same as the rules for model-building:

- beginning with our baseline measurements of performance at our point of origin (Level 1)

- aiming toward our culminating mission of our point of destination (Level 5)

- projecting our new processing threshold over which we must cross (Level 3)

- generating our developmental steps to approach this processing threshold (Level 2)

- collaborating on cumulative processing to accomplish the culminating mission (Level 4).

When we have achieved our scaling mission, we address the scaling of dimensions with which our scales may interact or interrelate.

Now we may view the scientific operations in sharp relief (see Figure 18):

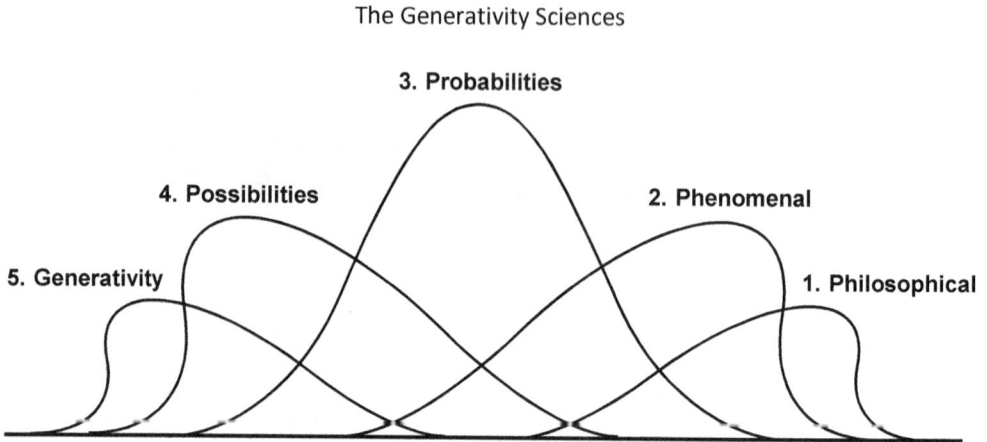

Figure 18. The Evolution of Scientific Operations

As each level of our paradigmetric scales have evolved developmentally and exponentially, each science has yielded to its higher-order counterparts. Together, they have been *true* to the mission of all science:

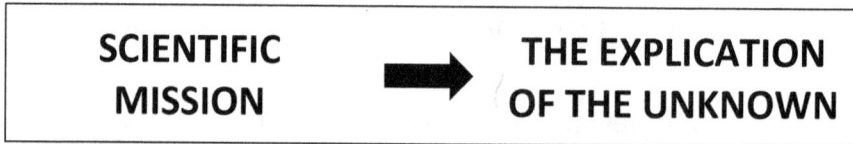

SCIENTIFIC MISSION ➡	THE EXPLICATION OF THE UNKNOWN

In transition, we began with implementation of the Sciences of Change: Probabilities Science, Possibilities Science, Generativity Science. There we crossed the threshold of Generativity Science by being introduced to Hypothetical–Deductive Model Building.

We transition into our understanding phase with an application of the fundamental factors of all science: human and information capital development (in an interdependent and synergistic relationship) dedicated to the New Science of Change Functions: describe, predict, relate, empower, free (see Figure 19).

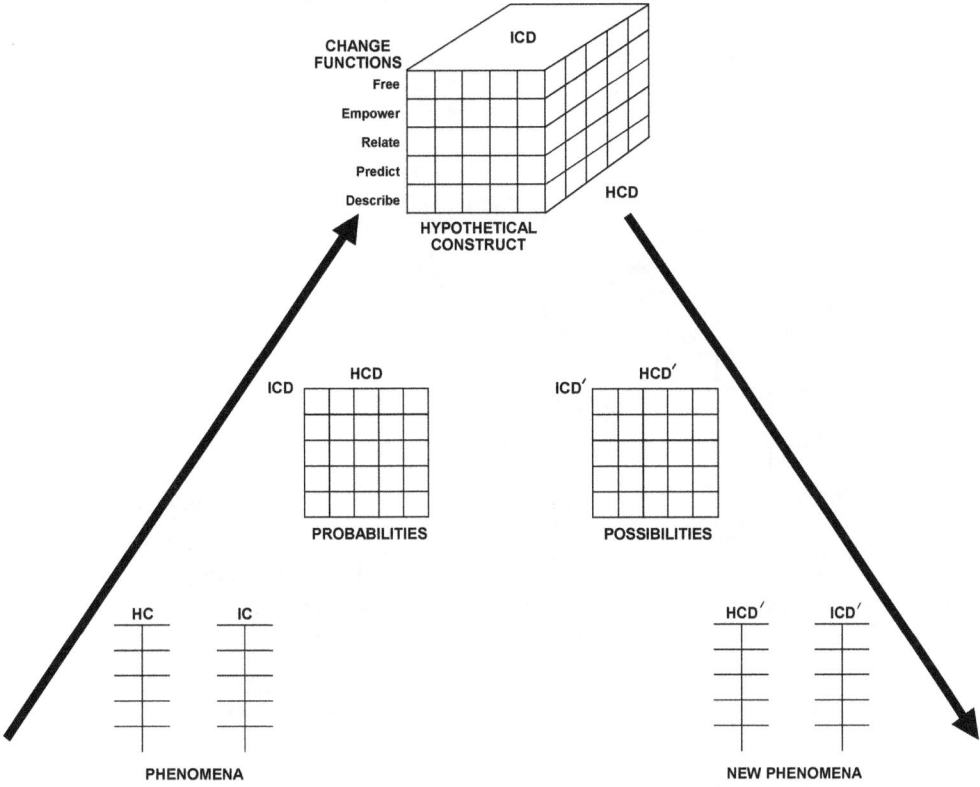

Figure 19. The Generation of the Science of Change

Now we can conclude our action phase with an image of the paradigm or model for the Science of Change (see Figure 20).

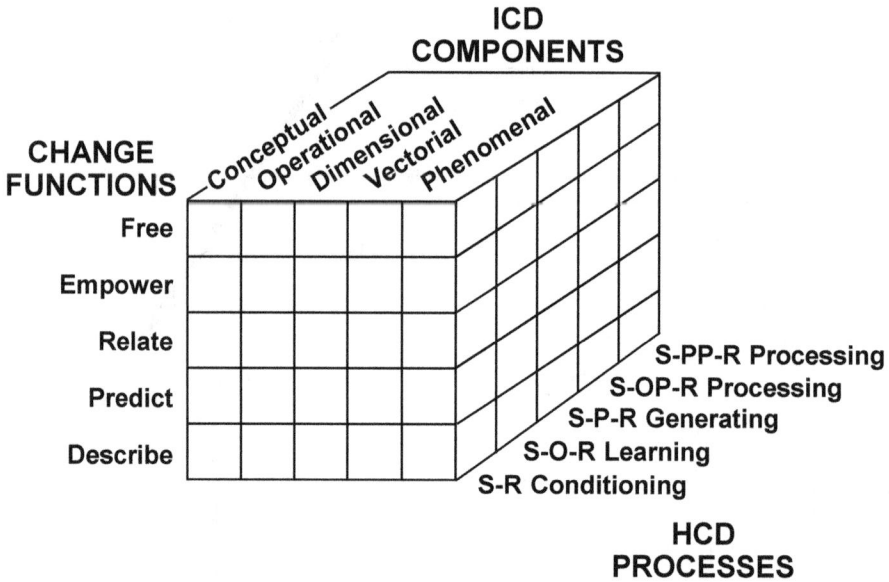

Figure 20. The Science of Change Model (Inductive)

Below are illustrated some examples of the differential benefits of operations under different socioenvironmental conditions following the format for verbal articulation of modeling:

Under	*Conditions*
If	*Components*
By	*Processes*
Then	*Functions*
So that	*Standards*

Probabilities Sciences

For example, here is what we find under conditions of authoritarian control in governance and economic enterprise:

Under	*Authoritarian Conditions (Top-Down)*
If	*Human and Information Potential*
By	*Are Processed by Probabilities Sciences*
Then	*Limited Human and Information Resources*
So that	*Performance and Productivity Remain Static*

Clearly, a static curve yields stagnant **GDP.**

Possibilities Sciences

For example, here is what we find under conditions of authoritarian control in governance and economic enterprise:

Under	*Mixed Authoritarian and Free Conditions (360°)*
If	*Human and Information Resources*
By	*Are Processed by Possibilities Science*
Then	*Expanded Human and Information Resource Development*
So that	*Performance and Productivity Grow Intentionally*

Clearly, openness to possibilities yields growthful **GDP.**

Generativity Sciences

Finally, here is what we generate under conditions of enlightened governance and entrepreneurial enterprise:

Under	*Free Conditions (360°)*
If	*Human and Information Resources*
By	*Are Processed by Generativity Sciences*
Then	*Human and Information Capital are Developed*
So that	*Performance and Productivity Grow Continuously*

Clearly, prosperity is continuous and explosive.

In summary, the **Science of Science** has evolved intentionally in the 21st century. It empowers us with the capacity to generate our own solutions to our own human problems. It is the source of the **Science of Change** (see Figure 21):

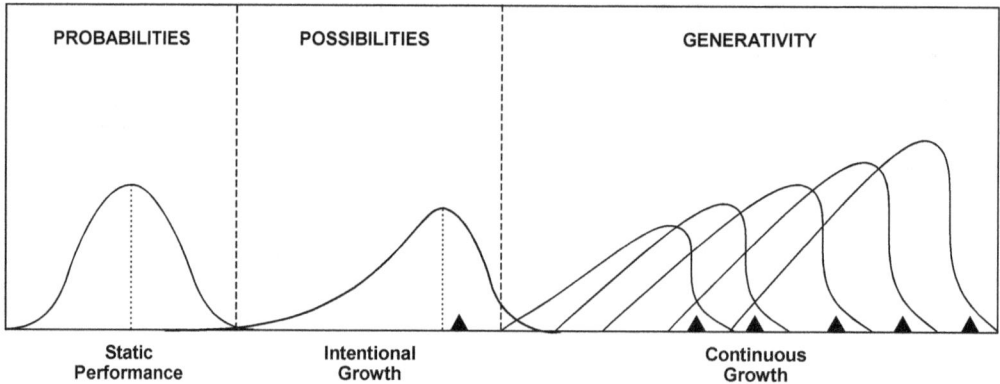

Figure 21. The Science of Change and Its Effects Upon Growth

We may view the sciences in their various modes in the Science of Change:

- **The Probabilities Sciences** describe characteristics or performance within normally distributed parameters, which empowers us to gain an understanding of individual performance relative to others in a static reference group.

- **The Possibilities Sciences** describe the processing of phenomena as they move toward incorporating exemplary performance, thus empowering us to model the processing of exemplars for purposes of intentional growth.

- **The Generativity Sciences** describe the continuous processing of all phenomena as they move continuously toward generating and encompassing exemplary performance related to their mission of continuous growth.

The choice is ours. We can experience the full benefits of our Science of Change. Or we can stagnate with our acquiescence to the processes and products of Probabilities Science! We can continue in our current malaise.

In transition, the Science of Science is available to empower us to select our own Science of Change:

- **Probabilities Sciences** to control our own predetermined destinies

- **Possibilities Sciences** to empower our own intentional destinies

- **Generativity Sciences** to generate our own changeable destinies

We may use these sciences sequentially when exploring new phenomena in order "to explicate the unknown."

When matured, we may employ these sciences simultaneously in integrated processing to generate "new phenomenal universes" because our current "universe" is unacceptable.

We can "ride a beam of light" in Einstein's terms.

Or we can "become beams of light" in Berenson's terms!

III

The Principles of
Generativity Sciences

7 Nesting, Encoding, Rotating— The Social Principles

There it was! Our formula for functional organizational systems:

$$\text{ODNA} \rightarrow n! \text{ (nest! encode! rotate!)}$$

Possibilities Sciences

This may be read as follows:

Organizational DNA is a function of n dimensions **"factorialized" ($n!$)** with all dimensions **"nested," "encoded,"** and **"rotated."**

This is our guiding formula for the operations of **Organizational DNA.**

It was the Fall of 1988 and we were boldly employing our newly minted Possibilities Sciences to generate a model for Organizational DNA. Working with my son, Christopher, I had engaged all of my mentors to assist in building the ODNA paradigm: Bernie Berenson, my life-long processing partner; B. R. Bugelski, my science and learning theory mentor; Jim Drasgow, my mentor for higher math and higher-order processing.

This was the formula that we were able to put together from processing interdependently with my colleagues. It was to lead to a powerful understanding of the universes of possibilities. Hopefully, over the course of this brief treatise, this will be made known to you.

What gave my colleagues great confidence was their understanding of Albert Einstein's theoretical equations reflecting the relationships of man and nature:

$$E = mc^2$$

In his second Relativity paper, Einstein made a historic deduction: If a body gives off an amount of energy (E) in the form of light, its mass will be reduced by that amount divided by the speed of light squared ($m = E/c^2$).

Einstein then made a great intellectual leap in transforming the algebraic equation: $E = mc^2$.

This new formula established that even a small amount of matter held the power of tons of explosives. This opened the door to the Nuclear Age. It also led to explanations for why the sun could burn for so many billions of years while not shrinking appreciably in size.

Einstein himself said:

> It is customary to express the equivalence of mass and energy (though somewhat inexactly) by the formula $E = mc^2$, in which c represents the velocity of light, about 186,000 miles per second; E is the energy that is contained in a stationary body; m is its mass. The energy that belongs to the mass m is equal to this mass, multiplied by the square of the enormous speed of light—which is to say, a vast amount of energy for every unit of mass.
>
> But if every gram of material contains this tremendous energy, why did it go so long unnoticed? The answer is simple enough: so long as none of the energy is given off externally, it cannot be observed. It is as though a man who is fabulously rich would never spend or give away a cent; no one could tell how rich he was. (Einstein 1956, pp. 51–52)

So it is with human beings. All are fabulously rich with potential. Few ever spend or give away a cent.

All human beings have the energy equivalent of tons of explosives—enough to energize many lifetimes—if they will but commit their physical resources.

All human beings have the emotional potential to expand their humanity to incorporate within their boundaries all manner of life throughout the ages—if they will but open to their emotional experience.

All human beings have the intellectual potential to grasp the wisdom of the ages while projecting their visions of the future. They can become what they will be—if they will but choose to use their intellects.

Yes, human beings can become masters of their own destinies, using their most precious gifts to shape or reshape their worlds as well as or, perhaps, better than those into which they were thrust. But to do so they must choose to live fully.

The most precious gift of all is the gift of intelligence. It is this gift that allows us to go places we have never been, to become people we have never known. It is this gift that allows us to throw a "skyhook" to our visions of what can be—energized by our physical resources, mobilized by our emotional resources, operationalized by our intellectual resources. It is this gift that enables us to "pull ourselves up by our own bootstraps" to our own ideals in fulfilling our visions of humanity.

A practical application of this gift may be viewed most fully with our success in landing a person on the moon. Pursuing his vision, NASA's Dr. John Houbolt operationalized the objective of a lunar-orbit rendezvous by lifting out of earth's orbit around the sun and descending down from the moon's orbit around the earth for the lunar-landing mission. Thus, he empowered us to "put a man on the moon"—a product of humankind's most vivid imagination!

Anything that can be conceived can be operationalized. It follows that anything that can be operationalized can be achieved. Achievement was purely a mechanical implementation process that launched and returned the spacecraft to fulfill the mission and the vision.

What gave us confidence in pursuing our deductive model for **Organizational DNA** was Einstein's courage in putting forth his statement of phenomenal possibilities. This put his principle of **Phenomenological Identification** within reach for us.

In this context, Einstein was also concerned with reversing the relationships in his equation. In so doing, he concluded that for the mass (**m**) increase to be measurable, the change of energy (**E**) per mass unit must be enormously large. He then introduced us to the terms that have become very familiar in our modern atomic world: **Radioactive Disintegration.** Schematically, the process is as follows: An atom of mass, **m**, split into two atoms of the mass, **m'** and **m''**, which separate with tremendous kinetic energy. If we can imagine these two masses as brought to rest, then together they constitute somewhat less energy than was the original atom—in contradiction of the old principle of the conservation of mass.

Einstein indicated that, while we could not actually weigh the atoms individually, there were indirect methods for measuring them. In addition, he postulated the kinetic energies that are transferred to the disintegration products, **m'** and **m''**. Thus, it became possible to test and confirm the equivalence formula and to calculate in advance—from precisely determined atom weights—just how much energy will be released with any atom disintegration.

What takes place can be illustrated with the help of our rich man. The atom **m** is a rich miser who, during his life, gives away no money (energy). But in his will, he bequeaths his fortune to his two sons **m'** and **m''**, on the condition that they give to the community a small amount, less than one thousandth of the whole estate (energy or mass). The sons together have somewhat less than the father had (the mass sum **m'** and **m''** is somewhat smaller than the mass **m** of the radioactive atom). But the part given to the community, though relatively small, is still so enormously large (considered as kinetic energy) that it brings with it a great threat of evil. Averting that threat has become the most urgent problem of our time.

Generativity Science

Let us conclude by catching up with our Organizational DNA Project. Remember its operational formula:

$$\text{ODNA} \rightarrow n! \text{ (nest! encode! rotate!)}$$

The core architecture for **ODNA** may be viewed in Figure 22:

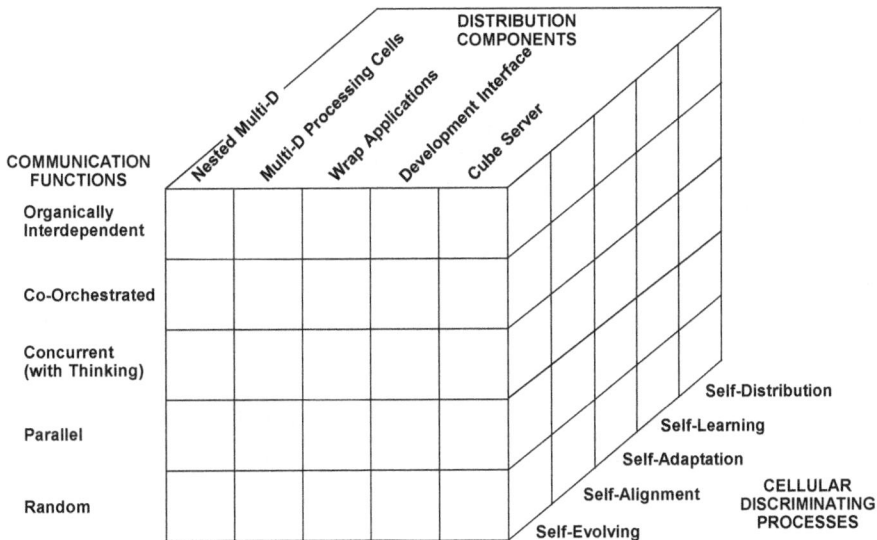

Figure 22. The Core Architecture for ODNA (Deductive Model)

The intersecting dimensions may be read as follows:

- **ODNA Communication Functions** range from "Random" to "Organically Interdependent."

- **ODNA Distribution Components** range from "Cube Server" to "Nested Multidimensional."

- **ODNA Cellular Discriminating Processes** range from "Self-Distribution" to "Self-Evolving Distribution."

The guiding architecture may be read as follows:

> Organically interdependent-driven communication functions are achieved by nested multidimensional distribution components empowered by self-evolving cellular discriminating processes.

The ODNA Platform has powerfully leveraged benefits as follows: storage, behavior, fault tolerance, high availability, self-healing, scaling, "hot code-swapping," authentication, meta processing, and easy deployment. In short, it is an IT power platform.

Guided by our formula of ODNA, then, we have now generated software platforms for mechanical, information, human, organization, and environmental technologies. Under the direction of Chris Carkhuff, we have validated the construct of cellular interdependency. Each of the billions of "smart cells" processes interdependently in multidimensional space to perform *all* of the organizational functions: from managing data, systems, and storage, to developing tools, applying middle-ware, and modeling business processes.

> Most powerfully, we have generated the multidimensional "cubing" platforms that enable all operating systems to communicate in any language to discharge any functions.

To sum, we have architected and implemented the most powerful **IT** platform:

Functions:	*Interdependent Processing Functions*
Components:	*Massively Concurrent Nested Multidimensional Components*
Processes:	*"Cubing" Platform Processes*

Verbally, this is the definition:

> Interdependent processing functions are accomplished by massively concurrent multidimensional components empowered by "cubing" platform processes.

This **interoperability** functions universally

- with all operating systems;
- in any computer language;
- for any organizational function.

In a **Cloud-driven** linear-processing, dependent-thinking world, the marketplace implications are earth-shaking for the following functions:

- Setting the processing platform standards
- Generating nested multidimensional spreadsheets
- Architecting the infinite web grid

In the process, we have generated the conditions for redefining all of the IT definitions of all human endeavors. In so doing, our IT platform has created an elevated platform for future civilizations.

Perhaps the greatest breakthrough is generational:

- We have built upon the Probabilities Science and Parametric Measurement inherited from Berenson, Bugelski, Sprinthall, and others.

- We have broken through our parametric boundaries with the Possibilities Science and Non-Parametric Measurement of Drasgow, Siegel, Tukey, and others.

- We have actualized the intentionality of Generativity Science and Paradigmetric Measurement of Chris Carkhuff and George Paley, and others.

Succinctly, in the span of three generations, we have generated **"The New Scientific Vision of the Science of Science:"**

- optimizing Probabilities Science
- generating Possibilities Science
- focusing Generativity Science

All of these processing sciences may be employed continuously and interdependently in their evolution of intentionality.

Social Modeling

In the pages that follow, we will illustrate the social principles of the Generativity Sciences in a learning module. In this module, the **Social Matrix** of the IT binary code processors will be nested, encoded, and rotated into the **Social Models** of the Generativity Science (see Figures 23 through 29).

The Social Matrix

According to the McKinsey Report (McKinsey, 2013), "Harnessing the ability to make any interaction or activity social—to influence actions, solve problems, and innovate, potentially creating new types of organizations that are not constrained by traditional boundaries." (See Figure 23.)

SOCIAL FUNCTIONS	SOCIAL COMPONENTS			
	Actuators	Sensors	Cloud	Mobile
• Crowd-Sourcing				
• Internal Networks				
• External Networks				
• Social Feature				
• Reimagine Organization				

Figure 23. The Components and Functions of the Social Matrix

Information "Cubing"

The net of our processing of phenomenal experience may be viewed in our **Information "Cubing"** phenomena: phenomenal functions, components, and processes (see Figure 24). As may be noted:

- Phenomenal functions are driven by the discovery of new phenomena.

- Human processes are driven by the phenomenal processing systems **(S–PP–R).**

- Information components are driven by phenomenal representations.

We may summarize the result of **Information "Cubing"** operationally (see Figures 25 through 27).

> **Phenomenal functions are achieved by information components empowered by human-information processing.**

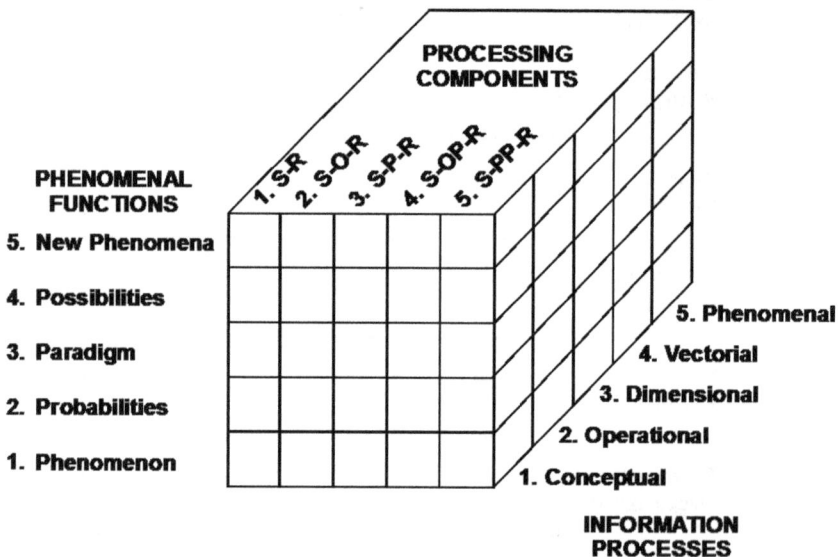

Figure 24. "Cubing" Phenomena

Information Levels

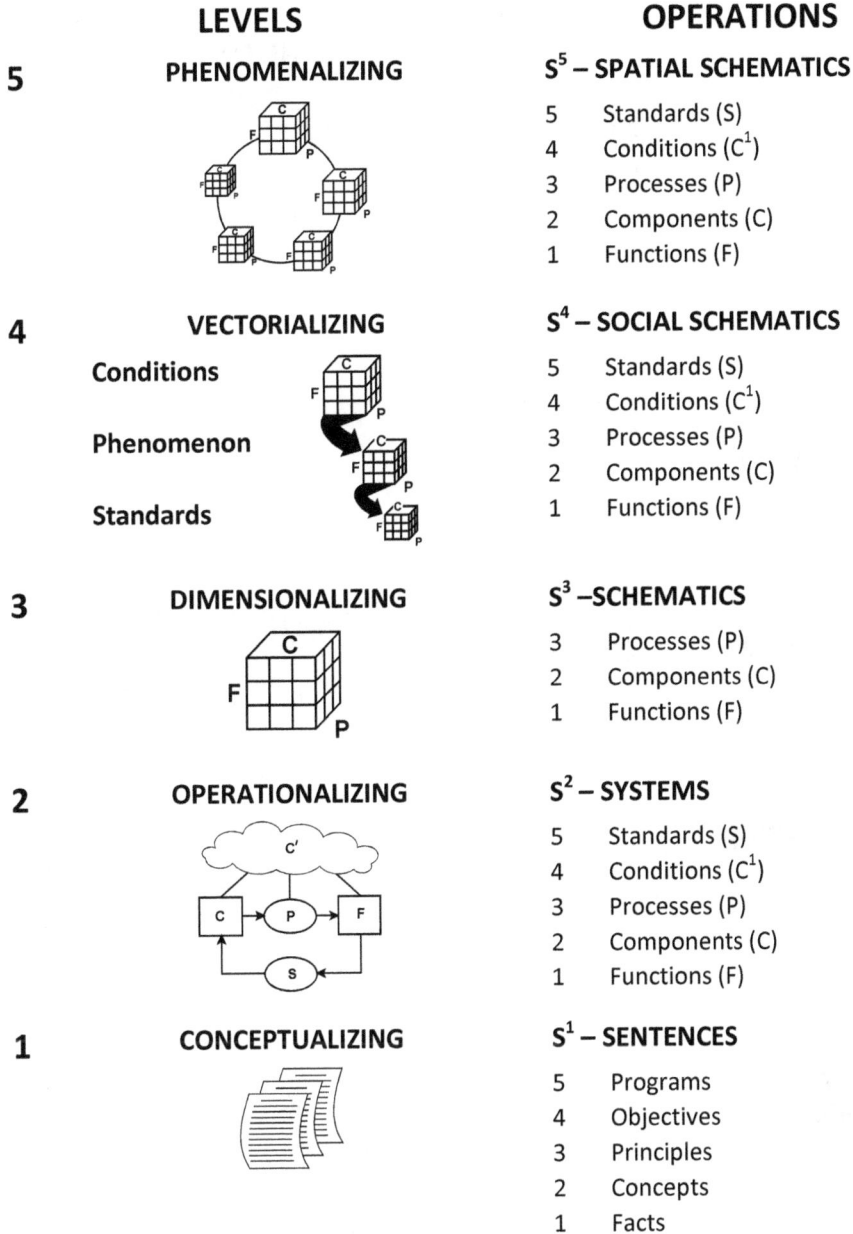

LEVELS

OPERATIONS

5   **PHENOMENALIZING**

S^5 – SPATIAL SCHEMATICS

5	Standards (S)
4	Conditions (C^1)
3	Processes (P)
2	Components (C)
1	Functions (F)

4   **VECTORIALIZING**

Conditions

Phenomenon

Standards

S^4 – SOCIAL SCHEMATICS

5	Standards (S)
4	Conditions (C^1)
3	Processes (P)
2	Components (C)
1	Functions (F)

3   **DIMENSIONALIZING**

S^3 –SCHEMATICS

3	Processes (P)
2	Components (C)
1	Functions (F)

2   **OPERATIONALIZING**

S^2 – SYSTEMS

5	Standards (S)
4	Conditions (C^1)
3	Processes (P)
2	Components (C)
1	Functions (F)

1   **CONCEPTUALIZING**

S^1 – SENTENCES

5	Programs
4	Objectives
3	Principles
2	Concepts
1	Facts

Figure 25. Levels of Information Representing (IR) Operations

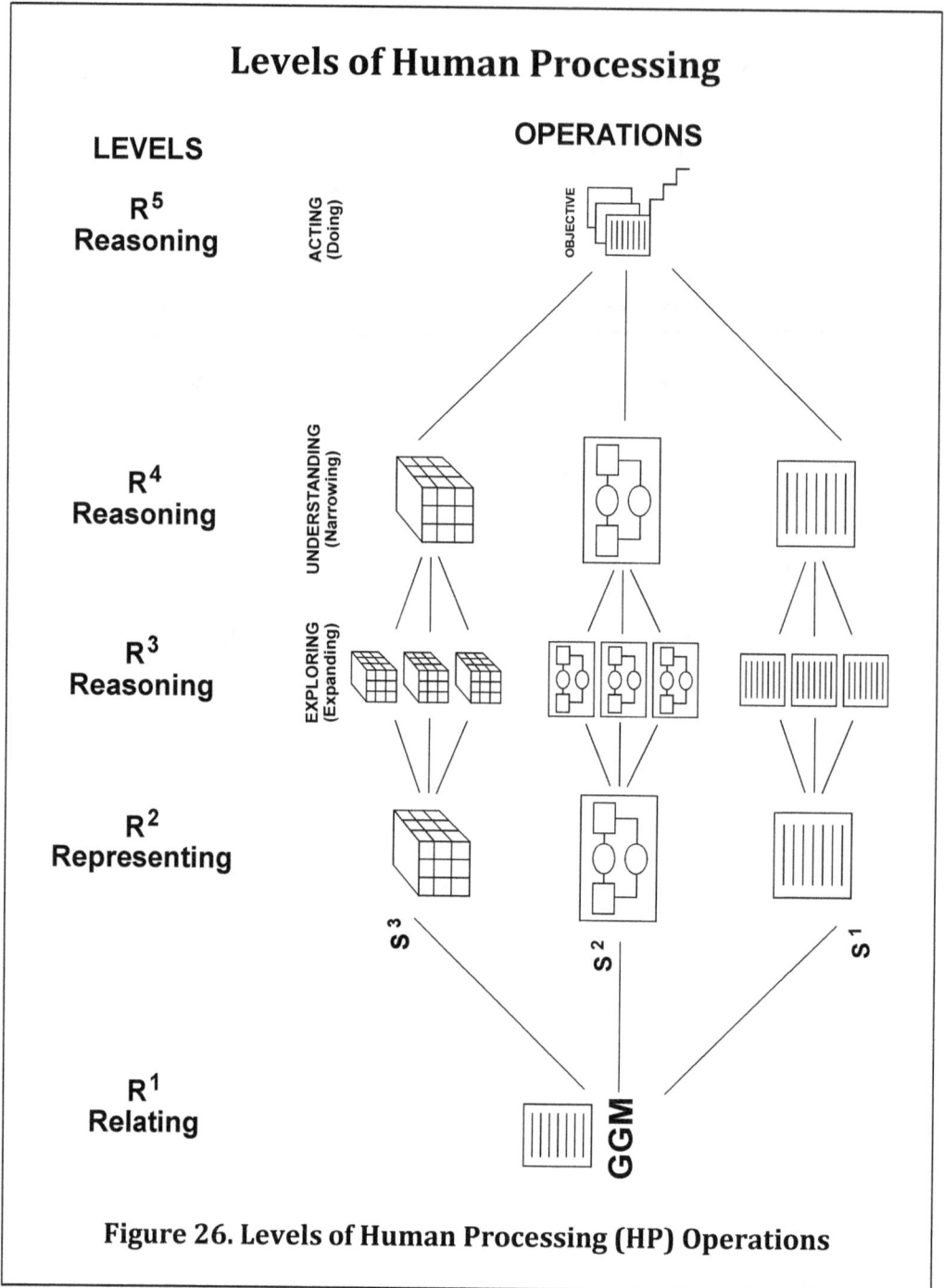

Figure 26. Levels of Human Processing (HP) Operations

Levels of Socioeconomic Organizational Systems

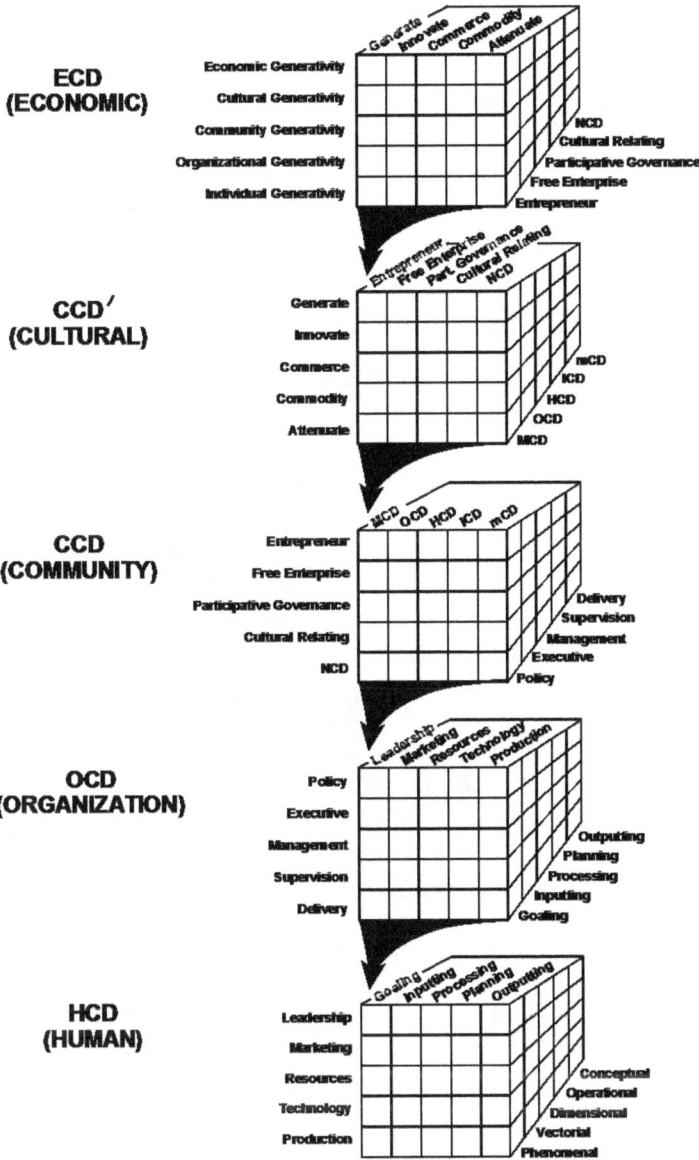

Figure 27. Socioeconomic Growth Systems

Social Modeling of the Social Matrix

The Social Matrix, itself, exists within the context or conditions of Social Modeling:

- dedicated to Human Generativity Functions
- transformed by Information Representation Components
- empowered by Information Technology Processes

We may view the impact of the different sciences in Social Modeling (see Figure 28):

- The Human Sciences upon human generativity functions

- The Information Sciences upon information representation components

- The Information Technologies upon social information technologies processes

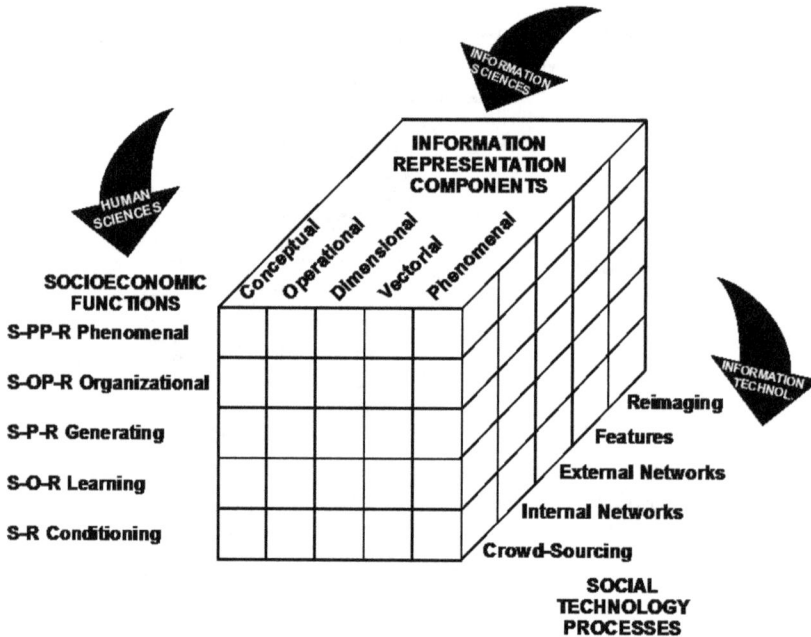

Figure 28. "Gen Cube" of the Social Matrix

The Social Model

The Social Model is the Core Phenomenal Model of social multidimensionality:

- Its socioeconomic functions are dedicated to economic prosperity and equity.

- Its generativity components emphasize the generative processing of the socioeconomic functions.

- Its information representation processes empower the generativity components to process the socioeconomic functions.

The Social Model derives its prepotency directly from the sciences:

- **The Basic Sciences—physical and natural**
- **The Human Sciences**
- **The Information Sciences**

Together, the dimensions of these sources define the mission of the **Social Model:** Socioeconomic functions are achieved by generativity components empowered by information processes.

The Social Model is the holy grail of all human and organizational endeavors (see Table 9).

Table 9. The Social Model

- **Basic Sciences**
- **Human Sciences**
- **Information Sciences**

We may view the **"Gen Cube"** of the Social Matrix into the Social Model in Figure 29.

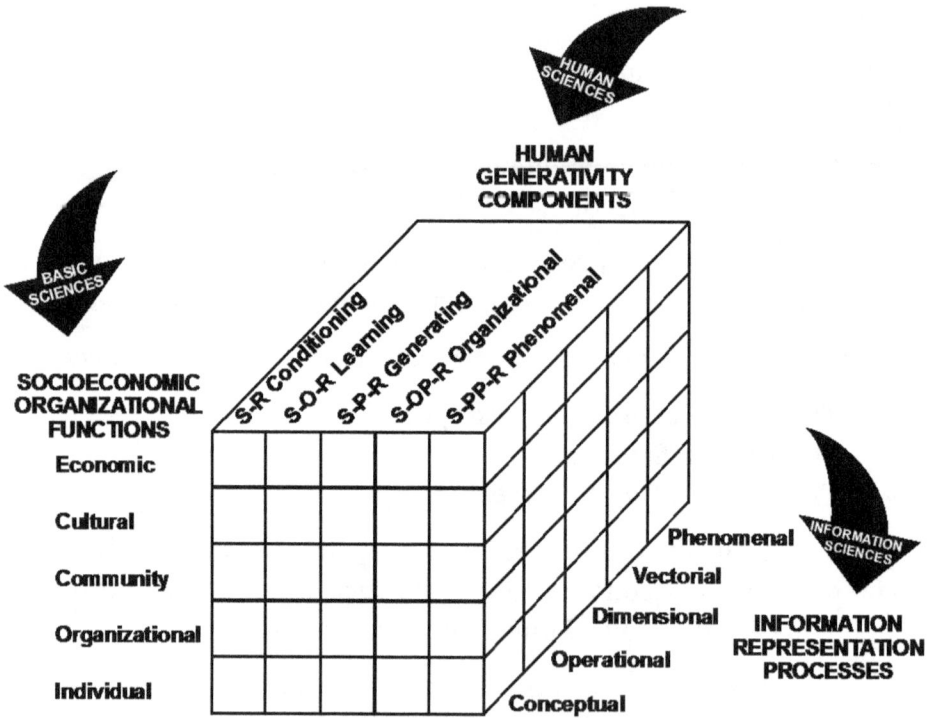

Figure 29. "Gen Cube" of the Social Matrix

Summary and Transition

In transition, what we have attempted to do in our interdependent efforts with Berenson, Bugelski, and Drawgow, and virtually with Einstein, da Vinci, and others, has been to accelerate the evolution of the sciences and, thus, their applications in civilization (see Table 10).

Table 10. Scaling the Operations and Measurement Systems of Levels of Science

Levels	Operations	Measurement
5	Generativity	Paradigmetrics
4	Possibilities	Non-parametrics
3	Probabilities	Parametrics
2	Phenomenal	Empirical
1	Philosophical	Anecdotal

As scientists, we have accelerated the movement of the evolution of science

- from the slow evolutionary changes of **Probabilities Science** and its controlling parametric functions;

- through **Possibilities Science** and its freeing non-parametric functions;

- to **Generativity Science** and its intentional paradigmetric functions.

We are at the same stage of development in understanding the actualization of our intellectual processing potential that Einstein was when he discovered the laws of physical energy. We are also, unfortunately, at the same stage of understanding the disintegration of society to support this.

The choice is ours. We can fulfill our human destiny by actualizing our intellectual potential or allow the disintegration of humanity by failing to energize and motivate ourselves to use our intellects. By choosing to actualize our potential, we show the courage to enter the unknown—confident in the possibilities that any goals that we can operationally define, we can achieve.

Clearly, humankind's goals in space or on earth are limited only by the boundaries of our intellect. Hopefully, we will not forego our home on earth prematurely. Hopefully, we will become "beams of light" here on Earth.

8 Phenomenal Processing: A Personal Note

It is a story told of the original "voyage of discovery" by Lewis and Clark. It is legendary that these explorers met different Indian tribes at different junctures of their voyage. And the "discovery" cohort made discriminations concerning their nature—good or bad.

The "good people," they believed, were a lot like themselves—warm and embracing, bright and alert. All around, they offered an elevated experience of what they would label "light" people with "bright" visions.

Today, we might label them "elevator" or "facilitator" variables in an equation of human endeavor.

The "bad people," in turn, they found to be a lot different from the way they viewed themselves—cold and distant, dark and disturbed. Altogether, they manifested a retarding experience of what they would label "dark" people with pathological visions.

Today, we would label them "retarding" or "depressor" variables in our equation for human endeavor.

What the "Discovery Voyages" did not know was the natural disposition of the native people. When confronting an "unknown" experience such as alien people coming from the East on ships, the natives adopted a "reflective" mode: they were disposed to "mirroring" the behaviors of the new people.

As such, they responded in different ways to different presenting behaviors:

- They were friendly with friendly visitors.
- They were unfriendly with unfriendly visitors.
- On some few occasions, they were even violently punitive with misbehaving visitors.

So the conclusions that the "discovery voyagers" drew were inferences from reflections of their own behaviors.

The Native Americans adopted the reflecting strategy as survival or protective positioning:

> You show me your intentions and we will show you ours.

It gave them a chance to size up the strangers and behave differently.

Initially, we may interpret this reflecting strategy as the primitive behavior of hunter–gatherers. That would be a profound mistake.

The reflecting strategy is the way "smart" humans—indeed animals— survive by exploring the unknown phenomena in their lives: to suspend judgment while exploring the phenomena and the experience they present.

It is also the way of the scientist in explicating the unknown:

- to understand the operations by which these phenomena exist

- to engage these phenomena in a process of communication

- to align with these phenomena in order to optimize their contributions

Phenomenal Processing

The scientific strategies for explicating the unknown may be summarized in five principles of phenomenal individual processing leading to interdependent processing:

1. In the manner of the Native American, "walking in their moccasins" enables us to explore viscerally the experience of the phenomena— human and otherwise.

2. Seeing the world through their eyes and sensing their experiences enables us to understand the objectives of the phenomenal experience.

3. Processing their experience with the functions of their minds and the processes of their brains enables us to plan and act upon their experience.

4. Communicating interpersonally and sharing mutually stimulating experiences enable us to merge with their experience.

5. Learning to live interdependently with the spirits or souls of these phenomena enables us to optimize our mutual contributions.

While some of these processes may appear "other-worldly," they empower us to enter and align with the experience of otherwise alien people or phenomena in order to optimize the contributions of all of us.

The principles of phenomenal processing prepares us for the five principles of generativity processing:

1. Initiating processing probabilistically in order to maximize description and prediction of phenomenal behavior.

2. Transforming possibilistically in order to optimize relating and empowering to generate new and improved images of phenomenal behavior.

3. Culminating generatively in order to freely generate a new series of images of phenomenal behavior.

4. Evolving intentionally in order to optimize the possibilistic series of images of phenomenal behavior available in a changing instance.

5. Revolving inentionally by replacing all devolving systems with accelerating revolutionary systems that improve the processes and outcomes of our intentionality dramatically.

Clearly, humankind's goals in space or on earth are limited only by the boundaries of our intellect. Hopefully, we will not forego our home on earth prematurely. Hopefully, we will become "beams of light," continuously generating our own changeable destinies here on earth.

Generativity Websites

- TheMcLeanProject.com
- CarkhuffGenerativityLibrary.com

TheMcLeanProject.com

I. PHYSICAL AND NATURAL SCIENCES

1. Kakovitch, T. S. *Terra Forma: Physical and Natural Science Solutions.* McLean, VA: McLean Project, 2012.

2. O'Hara, S. *Sustainability: Natural Science Solutions.* McLean, VA: McLean Project, 2013.

II. SCIENCE OF SCIENCE

1. Carkhuff, R. R. *The Science of Science.* McLean, VA: McLean Project, 2012.

2. Carkhuff, R. R. *The Generativity Sciences.* McLean, VA: McLean Project, 2015.

III. HUMAN EMPOWERMENT

1. Banks, G., Benoit, D., and Bergeson, T. *Generativity Education.* McLean, VA: McLean Project, 2007.

2. Benoit, D. and Carkhuff, R. R. *The New 3Rs: Generativity Processing.* McLean, VA: McLean Project, 2013.

IV. ORGANIZATIONAL OPTIMIZATION

1. Armstrong, P., Carkhuff, C. J., and Epps, L. *Dash Cube: Communication in Context.* McLean, VA: Gen Star LLC, 2015.

2. Carkhuff, C. J., Carkhuff, R. R., and Cohen, B. *The Possibilities Economy. New Capital Development and the Air Wars.* McLean, VA: McLean Project, 2005.

V. SOCIOECONOMIC GROWTH

1. Carkhuff, R. R., Banks, G., Griffin, A. H., and Sprinthall, R. C. *The Springfield Miracle: Interventions and Outcomes.* McLean, VA: McLean Project, 2014.

2. Kakovitch, T. S. and O'Hara, S. *Physics and the New Economy.* McLean, VA: McLean Project, 2014.

CarkhuffGenerativityLibrary.com

I. THE BASIC SCIENCES

1. Kakovitch, T. S. *The Fifth Force.* Amherst, MA: HRD Press, 2012.

2. Kakovitch, T. S. *Anthropogenics.* Amherst, MA: HRD Press, 2014.

3. Kakovitch, T. S. *Collegium.* Amherst, MA: HRD Press, 2013.

4. Kakovitch, T. S. and O'Hara, S. *Physics and the New Economy.* Amherst, MA: HRD Press, 2013.

5. O'Hara, S. *Sustainability.* Amherst, MA: HRD Press, 2015.

6. O'Hara, S. and Kakovitch, T. *Agricultural Sustainability and the New Economy.* Amherst, MA: HRD Press, 2015.

II. THE INFORMATION SCIENCES

1. Carkhuff, R. R. *Information Generativity: Volume I: The New Science of Information.* Amherst, MA: HRD Press, 2015.

2. Carkhuff, R. R. and Carkhuff, C. J. *Information Generativity: Volume II. The Generative Processing of Information.* Amherst, MA: HRD Press, 2015.

3. Carkhuff, R. R. *Information Generativity: Volume III: The Generative Designs of Information.* Amherst, MA: HRD Press, 2015.

4. Carkhuff, R. R. and Benoit, D. *Information Generativity: Volume IV. The Generative Tools of Ideation.* Amherst, MA: HRD Press, 2015.

5. Carkhuff, R. R. *The Generativity Sciences.* Amherst, MA: HRD Press, 2015.

III. THE HUMAN SCIENCES

1. Berenson, B. G. *The Human Sciences. Carkhuff and the Human Sciences.* Amherst, MA: HRD Press, 2014.

2. Carkhuff, R. R. *Human Generativity.* Amherst, MA: HRD Press, 2013.

3. Carkhuff, R. R. *The Human Sciences: Volume I. Possibilities, Probabilities, and Generativity Sciences.* Amherst, MA: HRD Press, 2013.

4. Carkhuff, R. R. *The Human Sciences: Volume II. Possibilities, Probabilities, and Generativity Technologies.* Amherst, MA: HRD Press, 2013.

5. Carkhuff, R. R. *The Human Sciences: Volume IV. Interpersonal Skills and Human Productivity.* Amherst, MA: HRD Press, 2013.

6. Carkhuff, R. R. and Carkhuff, C. J. *Human Generativity: Volume I. The Scientific Legacy.* Amherst, MA: HRD Press, 2014.

7. Carkhuff, R. R. and Carkhuff, C. J. *Human Generativity: Volume II. The Experimental Sciences.* Amherst, MA: HRD Press, 2014.

8. Carkhuff, R. R. and Carkhuff, C. J. *Human Generativity: Volume III. Generative Thinking Skills.* Amherst, MA: HRD Press, 2014.

9. Carkhuff, R. R. and Benoit, D. *Human Generativity: Volume IV. The New 3Rs: Relating, Representing, Reasoning.* Amherst, MA: HRD Press, 2014.

10. Carkhuff, R. R. *The Generativity Civilization.* Amherst, MA: HRD Press, 2015.

References

- Basic Sciences
- Information Sciences
- Human Sciences

Basic Sciences

Costanza, R., et al. The Value of the World's Ecosystems Services and Natural Capital. *Nature, 387,* May 15, 1997.

Daly, H. *Beyond Growth: The Economics of Sustainable Development.* Boston, MA: Beacon Press, 1997a.

Daly, H. *Beyond Growth: The Economics of Sustainable Development.* Boston, MA: Beacon Press, 1997b.

Georgescu-Roegen, N. Energy, Matter, and Economic Valuation: Where Do We Stand? in Daly, H., & Uman, A. (Eds.). *Energy, Economics, and the Environment.* Boulder, CO: Westview Press, 1981.

Gowdy, J. The Revolution in Welfare Economics and Its Implications for Environmental Valuation and Policy. *Land Economics, 80,* 239–257, 2004.

Gowdy, J., & O'Hara, S. *Economic Theory for Environmentalists.* Florida: St. Lucie Press, 1995.

Gowdy, J., & O'Hara, S. Weak Sustainability and Viable Technologies. Special Issue: Nicholas Georgescu-Roegen. *Ecological Economics, 22*(3), 239–247, 1997.

Hall, C., & Klitgaard, K. *Energy and the Wealth of Nations: Understanding the Biophysical Economy.* New York: Springer, 2012.

Herfindahl, O., & Kneese, A. *The Economic Theory of Natural Resources.* Columbus, OH: Charles E. Merill, 1974.

Higgs, H. *The Physiocrats* (1st ed.). London: MacMillan, 1987.

Jansson, A., Hammer, M., & Folke, C. (Eds.). *Investing in Natural Capital: An Ecological Economics Approach to Sustainability.* Washington, DC: Island Press, 1994.

Kakovitch, T. S. *The Fifth Force.* Amherst, MA: HRD Press, Inc., 2012.

Kakovitch, T. S. *Anthropogenics.* Amherst, MA: HRD Press, Inc., 2013.

Kakovitch, T. S., & O'Hara, S. *Physics and the New Economy.* Amherst, MA: HRD Press, Inc., 2014.

Kondratiev, N. *The Long Waves in Economic Life.* USA: Kessinger Publishing, 2010. Malthus, T. R. *Essay on the Principles of Population.* JM Dent & Sons, Ltd., 1973.

McKibben, B. *The Age of Missing Information.* New York: Random House, 1992.

Meadows, D., Randers, J., Meadows, D., & Behrens, W. *The Limits to Growth: A Report for the Club of Rome's Project on the Predicament of Mankind* (2nd ed.). Universe Publishing, 1972.

O'Hara, S. From Production to Sustainability: Considering the Whole Household. *Journal of Consumer Policy, 18*(4), 111–134, 1995.

O'Hara, S. Toward a Sustaining Production Theory. *Ecological Economics, 20*(2), 141–154, 1997.

O'Hara, S. Internalizing Economics: Sustainability Between Matter and Meaning. O'Brien, J. C. (Ed.). *Essays in Honor of Clement Allen Tisdell, Part IV, International Journal of Social Economics, 25*(2/3/4), 175–195, 1998.

O'Hara, S. Production in Context: The Concept of Sustaining Production, in Farley, J. (Ed.). *Festschrift for Herman Daly.* Burlington, VT: Vermont University Press, 2013.

O'Hara, S., & Kakovitch, T. *Sustainability.* Amherst, MA: HRD Press, Inc., 2015.

Pearce, D., & Atkinson, G. Capital Theory and the Measurement of Sustainable Development: An Indicator of Weak Sustainability. *Ecological Economics, 8,* 103–108, 1993.

Rees, W., Wackernagle, M., Testermale, P. *Our Ecological Footprint: Reducing Human Impact on the Earth.* New Catalyst's Bioregional Series No. 9. New York: New Society Publishers, 1996.

Schumpeter, J. *The Theory of Economic Development: An Inquiry into Profits, Capital, Credit, Interest, and the Business Cycle.* Cambridge, MA: Harvard University Press, 1955.

References

Schumpeter, J. *Business Cycles: A Theoretical, Historical, and Statistical Analysis of the Capitalist Process.* New York: Martino Publishers, 2005.

Sombart, W. *Economic Life in the Modern Age.* Stehr, N., & Grundman, R. (Eds.). USA: Transactions Publishers, 2001.

Vatn, A., & Bromley, D. Choices Without Prices Without Apologies. *Journal of Environmental Economics and Management, 26,* 126–148, 1994.

Victor, P. A. Indicators of Sustainable Development: Some Lessons from Capital Theory. *Ecological Economics, 4,* 191–213, 1991.

Information Sciences

Armstrong, M., Carkhuff, C. J., and Epps, L. *Dash-Cube: Communication in Context.* McLean, VA: McLean Project, 2015.

Avery, J. *Information Theory and Evolution.* Singapore: World Scientific, 2003.

Bairstow, J. *The Father of the Information Age.* New York: Laser Focus World, 2002.

Bateson, G. *Form, Substance, and Difference in Steps to an Ecology of the Mind.* Chicago, IL: University of Chicago, 1972.

Bekenstein, J. D. Information in the Holographic Universe. *Scientific American,* 2003.

Beynon-Davies, P. *Information Systems: An Introduction to Informatics in Organization.* London: Palgrave, 2002.

Beynon-Davies, P. *Business Information Systems* London: Palgrave, 2009.

Brown, J. S. and Duguid, P. *The Social Life of Information.* Boston, MA: Harvard Business School, 2002.

Casagrande, D. Information as Verb: Re-conceptualizing Information for Cognitive and Ecological Models. *Journal of Ecological Anthropology, 3,* 4–13, 1999.

Deloitte. *Tech Trends 2013: Elements of Postdigital.* New York: Deloitte Publications, 2013.

Dusenbery, D. B. *Sensory Ecology.* New York: Q. R. Freeman, 1992.

Floridi, L. *Information: A Very Short Introduction.* Oxford: Oxford University Press, 2010.

Gleick, J. *What Just Happened: A Chronicle from the Information Frontier.* New York: Pantheon, 2002.

Gleick, J. *The Information: A History, a Theory, a Flood.* New York: Pantheon, 2011.

Goonatilake, S. *The Evolution of Information.* London: Pinter, 1991.

Headrick, D. R. *When Information Came of Age.* Oxford: Oxford University Press, 2000.

Hoagland, M. and Dodson, B. *The Way Life Works.* New York: Random House, 1995.

Liu, A. *The Laws of Cool: Knowledge, Work and the Culture of Information.* Chicago, IL: University of Chicago, 2004.

McKinsey Global Institute. *Ten IT-Enabled Business Trends for the Decade Ahead.* Boston, MA: McKinsey Publications, 2013.

Pérez-Montoro, M. *The Phenomenon of Information.* Lanham, MD: Scarecrow, 2007.

Roederer, J. G. *Information and Its Role in Nature.* Berlin: Springor, 2005.

Seife, C. *Decoding the Universe.* New York: Viking, 2006.

Solymar, L. *Getting the Message: A History of Communication.* Oxford: Oxford University Press, 1999.

Stewart, T. *Wealth of Knowledge.* New York: Doubleday, 2001.

Vigo, R. Representational Information. *Information Sciences, 181,* 4847–4859, 2011.

Virilio, P. *The Information Bomb.* London: Verso, 2000.

Watson, J. *Behaviorism.* Chicago, IL: University of Chicago Press, 1930.

Wicker, S. B. and Kim, S. *Fundamentals of Codes, Graphs, and Iterative Decoding.* New York: Sprager, 2003.

Yockey, H. P. *Information Theory and the Origins of Life.* Cambridge: Cambridge University Press, 2005.

Young, P. *The Nature of Information.* Westport, CT: Greenwood, 1987.

Human Sciences

Anthony, W. *The Principles of Psychiatric Rehabilitation.* Baltimore, MD: University Park Press, 1979.

Aspy, D. N., and Roebuck, F. N. *Kids Don't Learn from People They Don't Like.* Amherst, MA: HRD Press, 1977.

Banks, G. *From Bondage through Prosperity: Finding the Freedom in Thinking.* Amherst, MA: HRD Press, 2013.

Berenson, B. G. *The Possibilities Mind.* Amherst, MA: HRD Press, 2001.

Berenson, B. G. *Carkhuff and The Human Sciences.* McLean, VA: The McLean Project, 2013.

Berenson, B. G. and Cannon, J. R. *The Science of Freedom.* Amherst, MA: HRD Press, 2006.

Berners-Lee, T. *Weaving the Web.* Britain: Orion Business, 1989, ISBN 0-7528-2090-7.

Bierman, R. *Toward Meeting Fundamental Human Service Needs.* Guelph, Ontario: Human Service Community, Inc., 1976.

Bugelski, B. R. *Psychology of Learning.* New York: Holt, Rinehart & Winston, 1956.

Bugelski, B. R. *Principles of Learning.* New York: Praeger, 1979.

Carkhuff, R. R. *Helping and Human Relations. Volumes I and II.* New York: Holt, Rinehart & Winston, 1969.

Carkhuff, R. R. *The Development of Human Resources.* New York: Holt, Rinehart & Winston, 1971.

Carkhuff, R. R. *The Promise of America.* Amherst, MA: HRD Press, 1974.

Carkhuff, R. R. *Toward Actualizing Human Potential.* Amherst, MA: HRD Press, 1981.

Carkhuff, R. R. *Sources of Human Productivity.* Amherst, MA: HRD Press, 1983.

Carkhuff, R. R. *The Exemplar.* Amherst, MA: HRD Press, 1984.

Carkhuff, R. R. *Human Processing and Human Productivity.* Amherst, MA: HRD Press, 1986.

Carkhuff, R. R. *The Age of the New Capitalism.* Amherst, MA: HRD Press, 1988.

Carkhuff, R. R. *Empowering.* Amherst, MA: HRD Press, 1989.

Carkhuff, R. R. *Human Possibilities.* Amherst, MA: HRD Press, 2000.

Carkhuff, R. R. *The Age of Ideation.* Amherst, MA: HRD Press, 2007.

Carkhuff, R. R. *The Art of Helping.* Ninth Edition. Amherst, MA: HRD Press, 2009.

Carkhuff, R. R. *Saving America: The Generativity Solution.* Amherst, MA: HRD Press, 2010.

Carkhuff, R. R. and Berenson, B. G. *The New Science of Possibilities. Volumes I and II.* Amherst, MA: HRD Press, 2000.

Carkhuff, R. R. and Berenson, B. G., et al. *The Possibilities Organization.* Amherst, MA: HRD Press, 2000.

Carkhuff, R. R. and Berenson, B. G., et al. *The Possibilities Leader.* Amherst, MA: HRD Press, 2000.

Carkhuff, R. R. and Berenson, B. G., et al. *The Freedom Doctrine.* Amherst, MA: HRD Press, 2003.

Carkhuff, R. R. and Berenson, B. G., et al. *Freedom-Building.* Amherst, MA: HRD Press, 2003.

Carkhuff, R. R. and Berenson, B. G., et al. *The Freedom Wars.* Amherst, MA: HRD Press, 2005.

Carkhuff, R. R. and Berenson, B. G., et al. *The Possibilities Economy.* Amherst, MA: HRD Press, 2006.

Drasgow, J. Eclipsing All Great Works. Foreword, *The Freedom Wars.* Amherst, MA: HRD Press, 2000.

Einstein, A. *Relativity: The Special and General Theory.* New York: Henry Holt, 1931.

Einstein, A. *The Evolution of Physics.* Cambridge: University of Cambridge, 1938.

References

Einstein, A. *Collected Papers of Albert Einstein.* Princeton, NJ: Princeton University Press, 1989.

Hebb, D. O. *The Organization of Behavior.* New York: John Wiley and Sons, 1949.

Hull, C. L. *Mathematics—Deductive Theory of Rote Learning.* New York: Appleton–Century–Crofts, 1940.

Hull, C. L. *Principles of Behavior.* New York: Appleton–Century–Crofts, 1943.

Hull, C. L. *A Behavior System.* New Haven, CT: Yale University Press, 1952.

Kakovitch, T. S. *The Fifth Force.* Amherst, MA: HRD Press, 2012.

Kakovitch, T. S. *Collegium.* Amherst, MA: HRD Press, 2012.

Kakovitch, T. S. *Anthropogenics.* Amherst, MA: HRD Press, 2013.

Kakovitch, T. S. and O'Hara, S. *Physics and the New Economy.* Amherst, MA: HRD Press, 2014.

Kerner, O., et al. National Advisory Commission. *Report on Civil Disorders.* NY: Bantam Books, 1968.

Kilby, J. *First Successful Demonstration of Integrating a Transition with Resistors and Capacitors on a Simple Semiconductor Chip Defining the Monolithic Idea.* Dallas, TX: Texas Instruments, September 12, 1958.

Pavlov, I. P. *Conditioned Reflexes.* Oxford: Oxford University Press, 1927.

Rogers, C. R. The Necessary and Sufficient Conditions of Therapeutic Personality Change. *Journal of Consulting Psychology,* 1957, 22, 95–103.

Siegel, S. *Nonparametric Statistics for the Behavioral Sciences.* Washington, DC: American Association for the Advancement of Science, 1959.

Sprinthall, R. C. *Basic Statistical Analysis.* Boston, MA: Allyn and Bacon, 2011.

Sprinthall, R. C. Psychenomics. Afterword in R. R. Carkhuff, *Saving America: The Generativity Solution.* Amherst, MA: HRD Press, 2010.

Sprinthall, R. C. *SPSS.* Boston, MA: Pearson Education, Inc., 2009.

Straus, E. *Phenomenology: Pure and Applied.* Pittsburgh: Duquesne University Press, 1964.

Truax, C. B. and Carkhuff, R. R. *Toward Effective Counseling and Therapy.* Chicago, IL: Aldine, 1967.

Watson, J. *Behaviorism.* Chicago, IL: University of Chicago Press, 1930.

Carkhuff Body of Work

Robert R. Carkhuff, Ph.D.

The Generativity Sciences

An Annotated Body of Work

CONTENTS

1. Introduction and Overview

Carkhuff was 15 years out of graduate studies in psychology when the world of science first took note of his scientific contributions in the study of helping and human relations.

Today, Carkhuff is the most powerful force in the history of science, representing the science and technology of **human generativity** and formulating the human sciences.

With a profound commitment to establishing a **true science of human behavior,** Carkhuff has been the path-finding leader for the past 50 years as his body of work will testify.

In 1978, in terms of frequency of citations in psychology, Carkhuff ranked among the leading contributors (see Table 3). Two volumes of *Helping and Human Relations* and one volume on *Counseling and Psychotherapy* led the listings.

Table 3.
Most-Cited Books in Clinical Psychology [1]

Bandura, A. *Principles of Behavior Modification,* 1969.

Carkhuff, R. R. *Helping and Human Relations, Volumes I and II,* 1969.

Kelly, G. A. *The Psychology of Personal Constructs,* 1955.

Truax, C. B. and Carkhuff, R. R. *Toward Effective Counseling and Psychotherapy,* 1967.

Fenichel, O. *The Psychoanalytic Theory of Neurosis,* 1945.

Freud, S. *Zur Geschichte der Psychoanalytischen Bewegung (On the History of the Psychoanalytic Movement),* 1914.

2. Helping and Human Relations

Carkhuff led the revolution of the helping professions from theoretical to operational treatment in the late 1960s. He and his associates defined the effective ingredients of helping in operational terms:

- *Toward Effective Counseling and Psychotherapy.* Chicago, IL: Aldine, 1967 (with C. B. Truax).

- *Sources of Gain in Counseling and Psychotherapy.* New York: Holt, Rinehart & Winston, 1967 (with B. G. Berenson).

- *Beyond Counseling and Therapy.* New York: Holt, Rinehart & Winston, 1967. Basically, the helping dimensions, such as empathy and respect, facilitated helpee exploration leading to indices of therapeutic personality change. The book with Truax (Aldine, 1967) was listed among the most-cited books in clinical psychology. [2]

Carkhuff extended these core findings to all helping and human relations. Simultaneously, he developed training programs for learning these operational skills in the practice of helping:

- *Helping and Human Relations. Volume I: Selection and Training.* New York: Holt, Rinehart & Winston, 1969

- *Helping and Human Relations. Volume II: Practice and Research.* New York: Holt, Rinehart & Winston, 1969

- *The Art of Helping.* Amherst, MA: HRD Press, 1971

Operationally, the helper's skills were refined to emphasize responding, personalizing, and initiating in order to facilitate the helpee's process, which was expanded to incorporate exploring, understanding, and acting, all of which led to therapeutic change. Now in its ninth edition, *The Art of Helping* has sold more than one million copies. In turn, the two volumes of *Helping and Human Relations* were among the most-cited books in social sciences. [3]

Along with his three books, Carkhuff was, himself, identified among the most-referenced social scientists by *The Institute of Scientific Information.* [4] [5] [6] Altogether, these works culminated the operationalization of previously theoretical processes.

3. Educational and Community Applications

In the 1970s and 1980s, Carkhuff transferred his findings to human and community development in the private as well as public sectors:

- *The Development of Human Resources: Education, Society and Social Action.* New York: Holt, Rinehart & Winston, 1971

- *Toward Actualizing Human Potential.* Amherst, MA: HRD Press, 1981

- *The Exemplar: The Exemplary Performer in the Age of Productivity.* Amherst, MA: HRD Press, 1983

- *Sources of Human Productivity.* Amherst, MA: HRD Press, 1984

Operationally, Carkhuff defined the productive human performer in the productive organizational system that, in turn, is defined in the productive community development system.

John T. Kelly, Director Emeritus, Advanced Systems Design, IBM, Inc., offered this evaluation of Carkhuff's work:

> **Carkhuff offers us a vision of the future. It is a vision of a great Age of Productivity, an age in which the human products and services are effectively increased so that all people can share. It is a vision of an age in which the resource inputs, natural and otherwise, are effectively invested so that no people are deprived of their birthrights.**
>
> (Kelly, Foreword, *Sources of Human Productivity,* 1984, p. xii)

During this period, Carkhuff and his associates launched a series of training products in teaching, training, and instructional systems design:

- *The Skills of Teaching Series, Volumes I–VI.* Amherst, MA: HRD Press, 1977–1981

- *Instructional Systems Design, Volumes I and II.* Amherst, MA: HRD Press, 1984

- *Training Delivery Skills, Volumes I and II.* Amherst, MA: HRD Press, 1984

Systematically, these works broke teaching and training down into four skill sets: interpersonal skills, content development skills, lesson planning skills, teaching delivery, and classroom management skills. The educational initiatives culminated in the issue of the journal, *Education,* dedicated to Carkhuff. [7]

4. Human and Organizational Processing

In the late 1980s, Carkhuff created systematic skills for human processing or thinking with applications for the development of **Human and Information Capital:**

- *Interpersonal Skills and Human Productivity.* Amherst, MA: HRD Press, 1984

- *Human Processing and Human Productivity.* Amherst, MA: HRD Press, 1986

- *The Age of the New Capitalism.* Amherst, MA: HRD Press, 1988

- *Empowering: The Creative Leader in the Age of the New Capitalism.* Amherst, MA: HRD Press, 1989

Carkhuff's mentor, B. R. Bugelski, a protégé of Clark Hull, one of the founders of American psychology, commented as follows:

> **This work rationalizes all of our efforts in learning theory and promises the culmination of psychology in a science of processing.**
>
> (Bugelski, Review, *Human Processing and Human Productivity,* 1984)

Carkhuff summarizes the research of the effects of individual, interpersonal, and organizational processing systems in hundreds of studies of more than 150,000 recipients. Basically, Carkhuff defined Human Capital Development or HCD as generative thinking. Carkhuff's work was reviewed by the distinguished social scientist, C. H. Patterson, University of Illinois:

> **This revolution has an important social significance also. I have stated elsewhere that the extent to which a society and its institutions, including its economic systems, facilitate**

the development of self-actualizing persons constitutes the criterion for the goodness of that society. To the extent that our society incorporates Carkhuff's system, it will become a better society for all its members.

(Patterson, Foreword, *Interpersonal Skills and Human Productivity*, 1983, p. 5)

5. Probabilities and Possibilities Sciences

In the 1990s and 2000s, Carkhuff, Berenson, and associates introduced the Science of Possibilities to drive the historical Science of Probabilities:

- *The New Science of Possibilities, Volume I. The Processing Science.* Amherst, MA: HRD Press, 2000
- *The New Science of Possibilities, Volume II. The Processing Technologies.* Amherst, MA: HRD Press, 2000
- *Human Possibilities.* Amherst, MA: HRD Press, 2000
- *The Possibilities Leader.* Amherst, MA: HRD Press, 2000
- *The Possibilities Organization.* Amherst, MA: HRD Press, 2000
- *The Possibilities Economy.* Amherst, MA: HRD Press, 2005

Dr. David N. Aspy, renowned scientist, educator, and protégé of Robert Oppenheimer, commented on Carkhuff's contribution to the advancement of civilization:

To support his views, Carkhuff does not simply offer up the science of science. He also presents the most exhaustive body of research and demonstration on relating and empowering ever presented in behavioral science. More-over, he engaged in the most advanced demonstrations of phenomenal potential, including human, ever attempted...

Together, these process-centric breakthroughs will lead us to a grand new Age of Enlightenment—The Age of Ideation, and in the process, The Science of Freedom.

(Aspy, *Window on the Universe, The Science of Freedom* 2007, p. 224)

6. New Capitalism and Freedom-Building

Eighteen months after having been introduced to Carkhuff's book, *"The Age of the New Capitalism,"* (1988), Pope John Paul II issued his Papal Encyclical, **"The New Capitalism."** In it, the Pope conceded the prepotent power of **human generativity** and committed himself to the **free economy** as the only alternative after the failure of Communism:

Can it perhaps be said that, after the failure of Communism, capitalism is the victorious social system, and that capitalism should be the goal of countries now making efforts to rebuild their economy and society?

If by "capitalism" is meant an economic system that recognizes the fundamental and positive role of business, the market, private property, and the resulting responsibility for the measure of production, as well as free human creativity in the economic sector, then the answer is certainly in the affirmative, even though it would perhaps be more appropriate to speak of a "business economy," "market economy," or simply "free economy." (John Paul II, **Centesimus Annus,** 42.1–42.2, in Miller, ***Encyclicals***)

Source: Carkhuff, R. R. *The Freedom Wars.* HRD Press, 2004

Early in the 21st century, Carkhuff and associates introduced the models and systems for freedom-building:

- *The Freedom Doctrine.* Amherst, MA: HRD Press, 2003
- *Freedom-Building.* Amherst, MA: HRD Press, 2004
- *The Freedom Wars.* Amherst, MA: HRD Press, 2004

In evaluating the contributions of Carkhuff to the advancement of civilization, John R. Cannon, Chief Executive Officer, Human Technology, Inc., summarizes Carkhuff's contributions [8] [9]:

> **Together, these works represent the contributions of The Possibilities Science to generating The Human and Ideational Sciences that define The Science of Freedom.**

In this context, there is nothing more powerful than the human brain enriched by possibilities experience... All of his books and all of his demonstrations, collectively from the earliest to the latest, are the products of this processing phenomenon, profound alternatives for individuals, organizations, communities, cultures, and nations; indeed, for The Global Village and its Marketplace. The Sources of Freedom are Possibilities!

(Cannon, Preface, *Science of Freedom,* 2007, p. xiii)

7. Generativity and the Human Sciences

Carkhuff's lifelong passion has been generativity or generative human processing. His recent work has focused on resolving the socioeconomic problems of our time through generative processing at all levels of community, culture, and economy:

- *The Generativity Solution, Building the New Economy.* Amherst, MA: HRD Press, 2009

- *The Generativity Solution, Volume III: Community Generativity.* Amherst, MA: HRD Press, 2009

- *The Generativity Solution, Volume IV: Cultural Generativity.* Amherst, MA: HRD Press, 2009

- *The Generativity Solution, Volume V: Economic Generativity.* Amherst, MA: HRD Press, 2009

Working with his colleagues, his body of work has culminated in a series introducing the **Human Sciences:**

- *Human Generativity: An Introduction to the Human Sciences.* Amherst, MA: HRD Press, 2013

- *The Human Sciences, Volume I: Probabilities, Possibilities, and Generativity Sciences.* Amherst, MA: HRD Press, 2013

- *The Human Sciences, Volume II: Probabilities, Possibilities, and Generativity Technologies.* Amherst, MA: HRD Press, 2013

- *The Human Sciences, Volume III: Carkhuff and Human Generativity.* Amherst, MA: HRD Press, 2013 (by B. G. Berenson)

Barry Cohen, Executive Vice President, Parametric Technology Corporation, summarizes Carkhuff's contributions as follows:

> **His body of research has constituted the foundation for revolutions in all areas of human endeavor: human, information, and organizational resource development: government, corporate and community development; cultural, national, and now global economic growth. In short, he has changed the world by making social science a "true science."**
>
> (Cohen, Foreword, *Generativity Solution,* 2009, p. ix)

Hernan Oyarzabal, Executive Director Emeritus, International Monetary Fund, summarizes Carkhuff's theory of the prepotency of generativity as follows:

> **It is indeed the "Generativity of Human Brainpower" and not the "Economic Theory of Stasis" that holds the interdependent, enlightened, and entrepreneurial keys to our Prosperous, Participatory, and Peaceful future.**
>
> (Oyarzabal, An Open Letter on the Economy, *Generativity Solution,* 2009, p. xiii)

8. Summary and Transition

In summary, Carkhuff's body of work has differentiated him from all others in the history of science. Primary among his contributions has been the operationalization and application of **possibilities science and generativity.**

Carkhuff's greatest contributions may lie ahead—the transfers of possibilities science to his current list of generativity projects in a troubled world. It is left to his lifelong colleague, Bernard G. Berenson, an Einsteinian scholar, to place Carkhuff's work in historical perspective:

> **Carkhuff's contributions to universal processing, alone, qualify him for leadership among the greatest scientists of history. His "nesting, encoding, and rotating of processing systems" are the core processes of Nature's Generativity. In creating The Human Sciences, Carkhuff belongs in the**

Pantheon of Science along with the works of DaVinci, Newton, and Einstein.

> (Berenson, B. G. We Can Be Beams of Light.
> Foreword, *Human Sciences, Volume II*
> Amherst, MA: HRD Press, 2013)

In this context, Carkhuff is dedicated to the values of Albert Einstein, his exemplar:

It is a very high goal: free and responsible development of the individual, so that he may place his powers freely and gladly in the service of all mankind.

> (Einstein, *An Ideal of Service to Our Fellow Man,*
> 1950, p. 59)

Where Carkhuff joined the list of established contributors to Psychology in 1978, he now stands alone as the dominant force defining the **Human, Information, and Organizational Sciences** in the 21st century science and technology.

Along with his associates, he has architected and implemented new platforms for the following:

- **Generative Human Sciences**
- **Human-Centric Information Sciences**
- **Human-Centric Organizational Sciences**
- **Human-Centric Economic Sciences**
- **Generative Civilizations**

In so doing, Carkhuff and his associates have outperformed the entire body of the social sciences and provided leadership for the declining body of the physical sciences.

His colleague, Dr. Tom Kakovitch, a physical scientist, has the final word on generativity:

Nature grows with generativity. That is why no one holds a monopoly over intelligence.

> (Kakovitch, *Collegium,* 2012)

[107]

[1] Garfield, E. *The 100 Books Most Cited by Social Scientists.* Number 37, Institute for Scientific Information.

[2] Garfield, E. *The 100 Books Most Cited by Social Scientists.* Number 37, Institute for Scientific Information.

[3] Garfield, E. *The 100 Books Most Cited by Social Scientists.* Number 37, Institute for Scientific Information.

[4] Garfield, E. *The 100 Books Most Cited by Social Scientists.* Number 45, Institute for Scientific Information.

[5] Endler, Rushtore, and Rogdeger. Productivity and Scholarly Impact. *American Psychologist,* Vol. 33, No. 12, 1062–1082.

[6] Heesacker, Heppner, and Rogers. Classics and Emerging Classics. *Journal of Counseling Psychology,* Vol. 29, No. 4, 400–406.

[7] Carkhuff, R. R. Leader in Human Resource Development. *Education,* Vol. 106, No. 3.

[8] Berenson, B. G. and Cannon, J. R. *The Science of Freedom.* McLean, VA: American Noble Prize, 2007.

[9] Carkhuff, R. R. *The Age of Ideation.* McLean, VA: American Noble Prize, 2007.

www.carkhuff.com

www.mcleanproject.com

www.carkhuffgenerativitylibrary.com